Hobby Hydroponics

Hobby Hydroponics

By Howard M. Resh, PhD

CRC Press is an imprint of the
Taylor & Francis Group, an **informa** business

CRC Press
Taylor & Francis Group
6000 Broken Sound Parkway NW, Suite 300
Boca Raton, FL 33487-2742

Printed in the United States of America on acid-free paper
15 14 13 12 11 10 9 8 7 6
International Standard Book Number-13: 978-0-931231-94-0 (Softcover)
Cover design by Sean Sciarrone

Library of Congress Cataloging-in-Publication Data

Resh, Howard M.
Hobby hydroponics / Howard M. Resh.
p. cm.
Includes bibliographical references and index.
ISBN 0-931231-94-9 (pbk. : alk. paper)
1. Hydroponics. I. Title.
SB126.5.R46 2003
631.5'85—dc21 2003042169

Visit the Taylor & Francis Web site at
http://www.taylorandfrancis.com

and the CRC Press Web site at
http://www.crcpress.com

Contents

List of Figures ix

List of Tables xv

Chapter One: Introduction/Background 1

Chapter Two: Starting Your Plants 9
- Seeds 9
- Sowing of Seeds in a Medium 14
- Hydroponic Culture in Choice of Medium 18
- Transplanting 18
- Seedling Temperatures 23
- Light for Seedlings 24
- Strawberries 24

Chapter Three: Cultural Practices 26
- Plant Spacing 26
- Watering 28
- Termperatures 28
- Light 29
- Types of Lights 30
- Amount of Light 31
- Reflectors & Wall Covering 31
- Ballasts 34
- Light Movers 35
- Carbon Dioxide Enrichment 37

Caring for Your Plants 38
Stringing 39
Training 41
Pollination 53
Planting Schedules 56
Crop Changeover 59
Controlling Pests & Diseases 60

Chapter Four: Plant Nutrition 66
Essential Elements 66
Soil in Comparison to Hydroponics 67
Sources of Essential Elements 68
Water Analysis 71
pH of Nutrient Solution 72
Electrical Conductivity (EC) of the Nutrient Solution 74
Symptoms of Nutritional & Other Problems 76

Chapter Five: Water Culture (True Hydroponic Systems) 79
Hydroponics vs. Soilless Culture 79
Water Culture Systems 80
Choosing the system 81
NFT Systems 82
A-Frame NFT 89
Ebb & Flow Water Culture Systems 92
Aeroponic Systems 94
Combination Water Culture & Soilless Systems 100

Chapter Six: Soilless Culture 114
Substrates (Media) 114
Soilless Systems—Expanded Clay 114
Rockwool Culture 123
Perlite Culture—Vertical Plant Towers 126
Hobby Plant Towers 133
Popular Hydroponics 136

Chapter 7: Hobby Hydroponic Supplies & Information 141
Hydroponic Supplies 141
Hydroponic Services 143
Hydroponic Organizations & Internet Chat Clubs 146
Hydroponic Magazines 148
References 149
Closing Comments 151

Index 154

List of Figures

Figure	Description
1.	Bucket system of gravel culture.
2.	Cross-section of a typical water-culture bed.
3.	"City Green" home hydroponic units.
4.	Components of an indoor unit. The use of air from an aquarium pump to move the nutrient solution up a tube to the growing tray.
5.	Inverted bottle in a nursery tray system for growing herbs and lettuce.
6.	Small home unit using a perlite-vermiculite medium.
7.	Seedling propagation cubes, blocks, peat-pellets and trays.
8.	Rockwool cubes, blocks and slabs.
9.	Tomato seedling in rockwool cube ready to transplant to rockwool block. Seedling is 15 days old. Note that the cube was laid on its side to permit the tomato plant to grow back upwards—allowing roots to grow along base of stem.
10.	Tomato seedlings spaced in tray and laid on sides.
11.	Pepper seedlings spaced in tray on their sides.
12.	Pepper seedlings transplanted to rockwool blocks and set in tray in a checkerboard pattern to utilize space efficiently.
13.	Cucumber seedlings (17 days old) growing in rockwool blocks spaced in a checkerboard pattern.
14.	Transplant European cucumbers to bato buckets at 3-week stage. Note the drip lines in the perlite medium and one in the block. Move the drip line from the

Figure	Description
	block to the medium after 7 to 10 days when the plant becomes established in the substrate.
15.	Bibb lettuce at 6" by 6" spacing in a Styrofoam board of a raft culture system.
16.	Parabolic light reflector.
17.	Horizontal light reflector.
18.	Conical light reflector.
19.	"Super Plus" horizontal lighting system.
20.	Ballast for a HID light.
21.	Three-arm circular light mover.
22.	"Sun Twist" two-arm, 180-degree rotating light system.
23.	Carbon dioxide generator that uses natural gas.
24.	"Tomahooks" for supporting plants.
25.	Plastic vine clip to support plants by attaching the hinge to the string and the clip underneath a healthy leaf.
26.	Side shoot of tomato plant must be removed early when about 1-inch long.
27.	"J" truss hook that attaches to support string and tomato truss to prevent the fruit truss from breaking.
28.	A plastic truss support that attaches directly to the fruit cluster.
29.	Pepper plants are trained with two stems.
30.	V-cordon system of training.
31.	Renewal umbrella system of training European cucumbers.
32.	Cucumber tendrils must be removed.
33.	Cucumber suckers (side shoots) must be removed.
34.	Small cucumber fruit about 1-inch long need to be removed from the plant up to 8 leaf axils to promote initial vegetative growth.
35.	Two laterals trained over the support cable in renewal umbrella training.
36.	Pollinating tomatoes using an electric toothbrush.
37.	Receptive flowers of tomatoes.
38.	Fruit set of tomatoes.
39.	Pruning basil early results in multi-stemmed plant. Note: place basil in top of plant tower pots.
40.	Bug-scan card to monitor insect populations.
41.	Beneficials available in shake bottles (*Encarsia-Eretmocerus* mix) or cardboard strips of pupae (*Encarsia formosa*).

Figure	Description

42. *Encarsia* pupae on strip is hung on a tomato leaf.
43. Cross-section of root with uptake of water and minerals from the nutrient solution into the vascular system.
44. Wide range of nutrients and other hydroponic supplies available at many hydroponic shops.
45. pH indicator paper strips.
46. Electrical conductivity (EC) meter.
47. Blossom-end-rot (BER) of tomato fruit.
48. Fruit cracking of tomato fruit.
49. Catfacing of tomato fruit.
50. American Agritech NFT unit "Jetfilm".
51. "Terrace Hydrogarden" by American Agritech.
52. "NFT Gully Kit" growing different lettuces.
53. Cross section of NFT channel.
54. "NFT Rockwool Gully Kit" by American Hydroponics.
55. "NFT Wall Garden" by American Hydroponics growing various types of lettuce.
56. "NFT Wall Garden" growing strawberries.
57. "612 NFT Production Unit" measures 6 ft by 12 ft with 9 production channels and one seedling channel. It is ideal for lettuce and herbs.
58. "Future Farms 196" grows 196 plants. Shown here is a combination of basil, bok choy and bibb lettuce on the left side and Red Sails and Green Salad Bowl leafy lettuces on the right side of the A-frame.
59. Corrugated plastic panels make up the sides of the A-frame where the nutrient solution flows back into the reservoir below.
60. The American Hydroponics "Vegi-Table" is an ebb-and-flow system that floods the plant roots from below and then drains back to a nutrient reservoir.
61. Healthy roots of plants growing in the ebb-and-flow tray of the "Vegi-Table". Note the fill/drain pipe in the base of the tray on the right.
62. "Pipe Dreams Balcony 18" aeroponic system.
63. "Pipe Dreams Balcony 32" aeroponic system.
64. "Pipe Dreams–PD96" aeroponic system.
65. "Pipe Dreams–PD160" aeroponic system is arranged in a "V" frame holds 160 plants.
66. The "Hobby Farm" of Diamond Lights is made of polystyrene plastic.
67. "Baby Bloomer" ebb-and-flow unit growing dwarf patio cherry tomatoes, basil and lettuce.

Figure	Description
68.	"AeroFlo^2" ebb-and-flow system by General Hydroponics.
69.	This small "AeroFlo^2" unit grows 42 low profile plants such as lettuce (lower left), arugula (center), spinach, etc.
70.	The "AeroFlo^2-20" has two 6-ft. growing chambers that hold 20 plants in 3-inch diameter net pots containing clay pellets or coco coir.
71.	"GroRox" clay pellets.
72.	The "AeroFlo^2-30" holds 30 plants in its three 6-ft. growing trays.
73.	The "AeroFlo^2-60" may be arranged with the six growing trays on one side of the reservoir for smaller plants as herbs and lettuce.
74.	The nice crop of lettuce growing in the "AeroFlo^2-60" system.
75.	Alternatively the "AeroFlo^2-60" may have 3 chambers on each side of the reservoir for taller plants such as tomatoes.
76.	"Micro Garden" ebb-and-flow system basic tray on top of a reservoir by American Agritech.
77.	Ebb-and-flow fill and drain pipes at the base of the growing tray of the "Micro Garden".
78.	Pots are filled with expanded clay pebbles in an ebb-and-flow system.
79.	Smaller net pots may be used instead of the larger square pots to hold the clay pebble substrate.
80.	A four-tray "Aerojet" aeroponic system.
81.	An "Aerojet" aeroponic system showing the mist "jets" in the base of the tray of the first tray. Note the nutrient reservoir below and the filter in the inlet line near the first tray.
82.	Four-tray aeroponic "Aerojet" system with four 6-inch diameter pots per tray.
83.	An aeroponic "Micro Garden" showing the mister jets at the base of the growing tray.
84.	Close-up view of mist nozzles and plumbing from pump below in the nutrient reservoir of the "Micro Garden".
85.	The aeroponic "Micro Garden" with the pots and plant support tops in place.
86.	The "Aero 9" internal view of mist system in the base of the growing tray sitting on top of the nutrient reservoir.

Figure	Description
87.	The same "Aero 9" unit with pots and tops closed.
88.	Diagram of "WaterFarm" one-pot system of General Hydroponics.
89.	"WaterFarm" components include pot, reservoir, pump, tubing and clay pellet substrate.
90.	"WaterFarm" may grow houseplants or vegetables.
91.	Components of "PowerGrower" system.
92.	Components of "AquaFarm" system.
93.	The "PowerGrower" may be connected in series to grow a number of plants. Nutrient reservoir is on the right.
94.	Diagram of the layout of "The Living System" by Light Manufacturing Co.
95.	A six-cell pot-on-pot unit of "Living System" requires a space of 5 ft by 5 ft.
96.	Cross-section diagram of "Dutch bato bucket" pot containing clay pellets.
97.	Diagram of "Eve's Garden" set up with Dutch pots on a supporting frame with drain channels, drip irrigation system and nutrient reservoir with pump.
98.	Six-pot system of "Eve's Garden".
99.	Vegetables growing in 12-pot system of "Eve's Garden". Zucchini, tomatoes, eggplant and melons.
100.	American Agritech "Econo-Jet" ebb-and-flow hydrogarden tray sits on top of a 20-gallon nutrient tank.
101.	A "Jetflo" model is similar to the "Econo-Jet" unit except that it is modular with several growing trays that may be set on a supporting frame above the nutrient reservoir.
102.	American Agritech "Jetstream" rockwool culture unit four-tray system. Note the use of the drip lines to each rockwool block.
103.	Six-tray top feed "Jetstream" rockwool culture unit.
104.	"Jetstream Mini" has 4 trays supported above a 10-gallon nutrient reservoir. It holds 8 plants.
105.	A "Jetflo Mini" uses a ebb-and-flow tray mounted to a 20-gallon reservoir. It holds 6 plants fed by a drip irrigation, rockwool culture system. These units are interchangeable for growing as ebb-and-flow or rockwool by some simple modifications using available conversion kits they term "AquaShuttle".
106.	Plant tower formed by stacking Styrofoam pots one on top of the other by turning them at 45 degrees to each other. Notches fit each pot in its exact position.

Figure	Description

107. Collection pot sits on drainpipe to retrieve all drainage. Swivel plate allows the plant tower to be turned easily each day to get uniform light on all of the plants.
108. Electrical conduit with plastic pipe sleeve supports pots vertically from an overhead support wire. Various herbs growing in plant towers of 10 pots.
109. Two drip lines enter the top pot and a third one is located in the middle of the plant tower with thyme.
110. Italian parsley growing in plant tower.
111. Chives in plant tower.
112. Thyme in plant tower. Note the plastic pipe sleeve on conduit support pipe at the top of the plant tower.
113. Bok Choy in plant tower.
114. Romaine lettuce in plant tower.
115. Mint, oregano and Italian parsley (right to left) in plant towers.
116. Two-stack, hobby plant tower. Basil planted in top pots, viola on left and marigolds on right.
117. Lettuce growing in three-stack, hobby plant tower. Note reservoir at bottom on which the plant towers sit. Pipe in center is to support the towers.
118. Irrigation bubbler at top of each plant tower stack. A timer-operated pump initiates irrigation cycles.
119. Bibb lettuce in left tower, beet tops in middle and oak-leaf lettuce in right tower.
120. Inexpensive sack culture of growing lettuce, herbs and strawberries in poor communities of Peru.
121. Grasses growing for cattle in inexpensive containerized hydroponics containing mixtures of rock, sand, peat, rice hulls and coco coir. This is "Popular hydroponics" in Peru.
122. Sack culture in Peru using a mixture of coco coir and rice hulls. Note collection pan at bottom and nutrient reservoir at top of each sack.
123. Inexpensive raft culture system for growing lettuce and herbs in poor communities of Peru.
124. Small 10½-ft. by 16-ft. backyard greenhouse.
125. Display of various growing systems in a hydroponic store.

List of Tables

Table	Description
1.	Seed varieties and sources.
2.	A general macronutrient formulation for a 10-U.S.-gallon tank.
3.	A 300-times strength micronutrient stock solution in 10-U.S.-gallon tank.
4.	Web sites of manufacturers of hydroponic units and accessories.
5.	Web sites of government and universities offering information on hydroponics.
6.	Web sites on identification and control of pests and diseases using integrated pest management (IPM).
7.	Web sites of seed houses.
8.	Addresses and Web sites/e-mail for hydroponic societies.
9.	Web sites/e-mails of hydroponic forums.
10.	Addresses and Web sites of hydroponic magazines.

1 Introduction/ Background

The objective of this book is to provide the reader with information on the basics of hydroponics that can be applied to a small-scale or hobby setup. I shall describe how to start your plants, to care for them and to choose from various hydroponic hobby units that may be most suitable for you. Details of the various hydroponic cultures and how you may construct them yourself or from where you may purchase ready-to-start units along with supplies and components are presented.

While the commercialization of hobby or "popular" hydroponics really became very active in the past 15 to 20 years, simple, self-designed systems were in use since the 1940's and 1950's. At that time the most common methods were gravel culture and water culture.

The bucket system of gravel culture was the most common where a small bed of gravel had a bucket attached to a hose joining them (Fig. 1). A bucket was filled with nutrient solution and then supported above the growing bed to allow the nutrient solution to flow into the bed flooding it from the bottom upward. This was a sub-irrigation gravel culture system. Within a period of 5 minutes when the bucket completely drains into the substrate void spaces of the bed, it is lowered

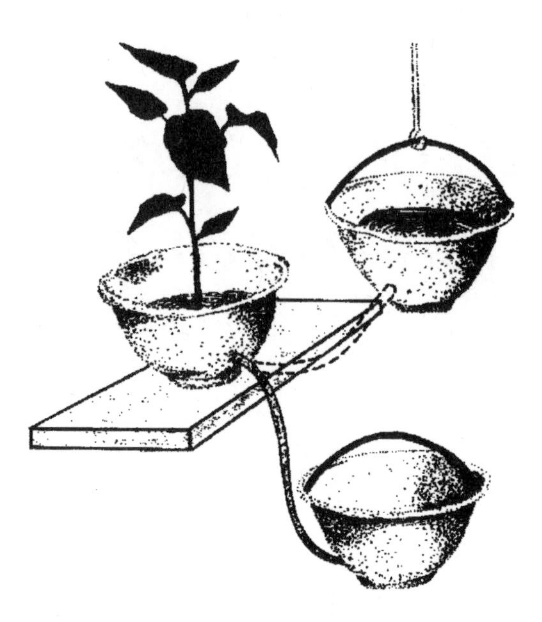

Figure 1. Bucket system of gravel culture.

below the level of the bed and the nutrient solution drains back to the bucket. This is repeated by hand a number of times a day depending upon the water demand of the plants that is determined by the temperature, light and stage of plant growth. While the system is very simple and works well, it requires that the hobbyist spend time during the day caring for his plants' irrigation needs.

A water culture system termed a "litter tray" system used a reservoir of nutrient solution in a rectangular tank with a tray of substrate located above the nutrient solution (Fig. 2). The most common medium was excelsior wood fibers, wood shavings, sawdust, dry straw, rice hulls or peanut hulls. The tray was 2 to 4 inches thick, constructed of wood with a wire mesh on the bottom to hold the substrate. Galvanized chicken wire of 1-inch diameter was coated with asphalt paint to prevent the release of zinc from its galvanized coating. The nutrient tank had a depth of 4 to 6 inches. The substrate would be watered when transplanting by hand for several days until the roots began growing

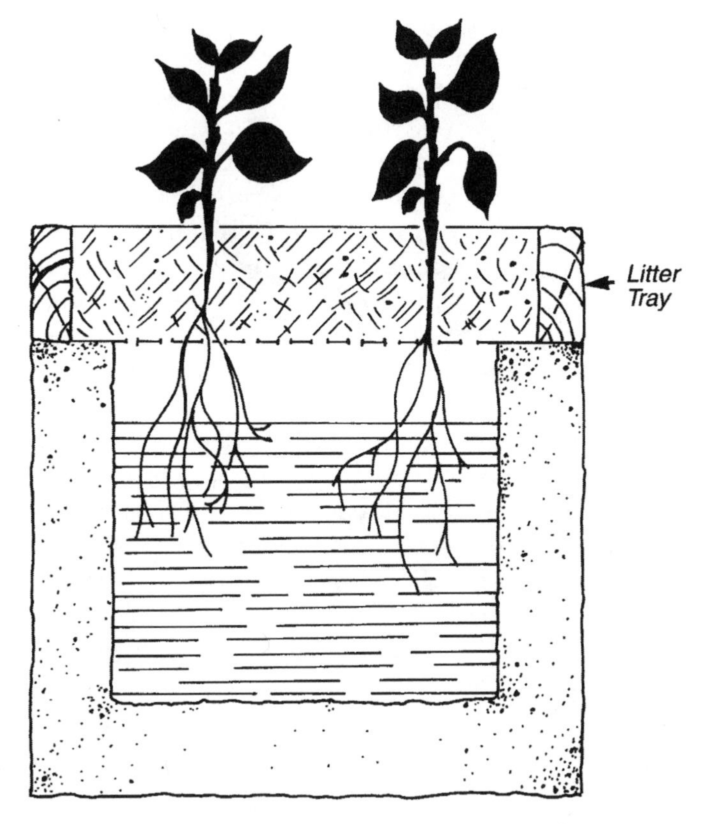

Figure 2. Cross-section of a typical water-culture bed.

into the nutrient solution below. Once the roots of the plants became established into the nutrient solution below, the solution level would be lowered gradually from 1 inch to 2 to 3 inches between the top of the solution and the base of the litter tray. This helped oxygenate the plant roots.

With the introduction of plastics, small pumps, timers and drip irrigation supplies, these similar designs could be modified to operate automatically. For example, using the principle of the litter tray, construct the tray of plastic sitting on top of a rigid plastic or fiberglass nutrient reservoir. The growing tray contains 2 to 4 inches of a substrate such as light-weight volcanic rock, Leca-clay pellets, Heydite-porous shale rock, perlite, sawdust, bark, or a mixture of rice hulls, peat or coco coir. Small holes of about 1/4-inch diameter are drilled in the bottom of the tray. A plastic screen is placed on the bottom of the tray to prevent the substrate from falling through into the nutrient tank below. A small fountain pump plugged into a simple household time clock will operate pre-set irrigation cycles. The pump is attached to a half-inch diameter black polyethylene hose containing trickle tubes with irrigation emitters as it rises on top of the growing tray. The excess nutrient solution percolates through the perforated bottom of the growing tray back to the nutrient reservoir below.

One of the earlier automated hobby units was the "City Green" hydroponicums (Fig. 3). These units were available in the 1970's. They were the first commercial attempt at small hobby hydroponic systems. They were constructed of molded plastic having a nutrient reservoir in the bottom and an upper growing tray. Expanded clay or volcanic cinder rock was the choice of growing medium. A fish-aquarium air pump was placed in one corner of the growing tray where it was attached to a small polyethylene tube that entered the

Figure 3. "City Green" home hydroponic units.

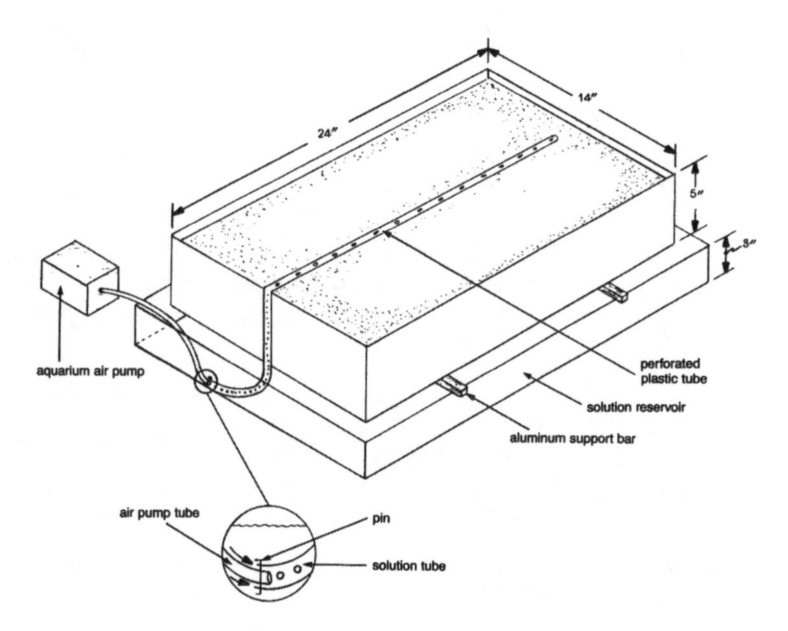

Figure 4. Components of an indoor unit. The use of air from an aquarium pump to move the nutrient solution up a tube to the growing tray.

nutrient tank below. This tube was inserted about one-half inch into a slightly larger diameter tube allowing a small space between the walls of the two tubes (Fig. 4). A pin held them together. As the aquarium pump forced air from the smaller tube into the larger one, the nutrient solution would be sucked into the tube together with the air bubbles. The solution with the air bubbles would rise in the larger tube to the surface of the growing tray where the tube was perforated to permit the nutrient solution to escape along with the air. The nutrient solution percolates through the medium and back to the reservoir underneath through the perforated bottom of the growing tray.

Perhaps one of the simplest hydroponic systems is the inverted bottle in a tray (Fig. 5). For the solution tray use a plastic flat of dimensions 10½ by 21 inches that has no holes. Place a bedding tray of 24 or 36 compartments in the flat, but remove one corner of the

Figure 5. Inverted bottle in a nursery tray system for growing herbs and lettuce.

filler tray to allow room for placement of a 1-gallon plastic jar. It must have a large plastic lid. Drill a 1/4-inch diameter hole in the middle of the large cap and glue a split cork ring of 3 inches in diameter on the cap. The small gap will allow the flow of nutrients from the bottle into the tray as the plants take up the solution. The bottle is inverted in the bottom of the tray. As solution flows to the plants, the level is maintained by air entering the bottle through the hole in the lid permitting a small amount of solution to flow from the bottle.

Use vermiculite or perlite as a substrate. You can seed directly into the medium. Water the seeds for several days until germination occurs before placing the inverted solution reservoir in the tray. You can cover the tray with plastic for several days until germination starts, then immediately remove the plastic or the seedlings will get long and skinny from excess heat. This tray is good for baby lettuce and herbs or a combination of lettuce, beets, herbs, upland cress, arugula, mustards, mizuna, orach, chard, and spinach to form a mesclun mix.

Over the past 20 years with increasing interest in hobby hydroponics, many small-scale units have been developed to meet market demand (Fig. 6). Now it is very easy to visit one of over 750 hydroponic stores worldwide to purchase hobby units and all the supplies you will need to get started. It is estimated that today over 1 million households in the United States operate small hobby hydroponic systems. This, of course, has spread to almost all countries of the world. While I do not have the figures for other areas, we know by the number of hydroponic stores that exist in other countries that they must have many hobbyists using household units.

The objective of this book is to make you aware of the types of hydroponic hobby units that are available on the market today and the supplies you need to get started. I shall describe in detail some of the units available and suggest which crops they are most suitable to growing. This information I hope will assist you in your decision to enter into hobby hydroponics.

Hobby hydroponics will provide you with the benefits of pleasure, rewarding products, clean products and relaxation. You will achieve self-satisfaction by growing

Figure 6. Small home unit using a perlite-vermiculite medium.

your own "garden fresh" salads. These salads will be highly nutritious and free of pesticides. Nutritional analyses of hydroponic tomatoes and peppers have demonstrated increases of up to 50% in vitamin and mineral content. These included vitamins A, B1 (thiamin), B2 (riboflavin), B3 (niacin), B6 (pyridoxine), C and E. By using bioagents and natural pest control measures your product will be free of synthetic pesticides. You will also feel relaxed and relieved from everyday working stress as you attend your plants in the hydroponic garden. It will allow your mind to escape from your daily concerns as you train and care for your plants. This is especially helpful during the dark days of winter as you are looking after your plants under supplementary lighting.

Another aspect of the hobby is that there are many Web sites, links and e-mail addresses on the Internet to get assistance and new ideas for your hydroponic growing. You may ask questions or just wish to "chat" with other hobbyists growing hydroponically. There are hydroponic associations in most countries that have annual conferences and regular meetings in which you may participate to learn new things and meet new friends having similar interests as yourself. I describe some of these in the last chapter.

2 *Starting Your Plants*

Seeds

The best method of starting your plants is from seed. The choice of variety of any plant is important. You should use those varieties that have been proven to grow best under hydroponic culture. While all varieties will grow well hydroponically, some special greenhouse varieties developed for controlled environmental conditions, such as you will have in your house or a hobby greenhouse, grow faster and yield higher than conventional field varieties. I have listed in Table 1 some varieties of lettuce, herbs, bok choy, tomatoes, peppers, and European cucumbers that I have found do well in hydroponic culture. This table is only a guideline, as there are many other varieties available that may produce well under your conditions. Feel free to test them yourself.

Table 1. Varieties and Sources

Plant	Variety	Source	Notes
Arugula	Astro II	Stokes Seeds	

Table 1. Varieties and Sources *(cont'd)*

Plant	Variety	Source	Notes
Bok Choy	Green Fortune	Ornamental Edibles	
Cucumber	Dominica	De Ruiter Seeds	Mildew Resistant
	Marillo	"	"
	Millagon	"	
	Discover	"	
	Corona	"	
	Accolade	"	
	Crusade	"	
	Logica	"	
Herbs	Sweet Italian Basil	Stokes Seeds, Richters, Johnny's Selected Seeds	All do well
	Purple Basil		
	Cinnamon Basil		
	Thai Basil	"	
	Lemon Basil	"	
	Chervil	"	
	Chives	"	
	Cilantro (Coriander)	"	
	Dill-Fernleaf		

Table 1. Varieties and Sources

Plant	Variety	Source	Notes
	Spearmint	"	
	Lavender-Lady	"	
	Oregano-Greek	"	
	Parsley-Curled	"	
	Parsley-Italian	"	
	Sage	"	
	Sorrel	"	
	Summer Savory	"	
	Sweet Marjoram	"	
	Thyme	"	
Lettuce	Bakito	Stokes Seeds	Romaine/Cos
	Brunia	"	Oakleaf
	Cimmaron	"	Romaine/Cos
	Freckles	Johnny's Seeds	"
	Ibis	Stokes Seeds	Red Looseleaf
	Lolla Rossa	Stokes Seeds	Red Looseleaf
	New Red Fire	"	"
	Rex	Rijk Zwaan	Bibb/ Buttercrunch

Table 1. Varieties and Sources (cont'd)

Plant	Variety	Source	Notes
	Red Sails	Stokes Seeds	Red Looseleaf
	Red Salad Bowl	"	Oakleaf
	Green Salad Bowl	"	"
	Vegas	Rijk Zwaan	Bibb/ Buttercrunch
	Waldmann's Dark Green	Stokes Seeds, Johnny's Seeds	Grand Rapids
	Black Seed Simpson		"
	Ostinata	Rijk Zwaan	Bibb/ Buttercrunch
Peppers	Blue Jay	Stokes Seeds	Purple/Red
	Dove	"	Ivory/Red
	Oriole	"	Green/ Orange
	Canary	"	Green/Yellow
	Cubico	De Ruiter Seeds	Green/Red
	Fellini	"	Green/ Orange
	Narobi	"	Green/ Orange
	Kelvin	"	Green/Yellow
	Samantha	"	Green/Yellow

Table 1. Varieties and Sources

Plant	Variety	Source	Notes
Tomatoes	Trust	De Ruiter Seeds	Beefsteak
	Match	"	"
	Grace	"	"
	Blitz	"	"
	Style	"	"
	Caruso	"	"
	Laura	"	"
	Perfecto	"	"
	Conchita	"	Cherry
	Favorita	"	"
	Dasher	"	Grape Cherry
	Picolino	"	Red Cocktail
	Flavorino	"	Plum Cocktail
	Bolzano	"	Orange Truss
	Locarno	"	Yellow Truss
	DRK 902	"	Orange Truss
	Tradiro	"	Red Truss
	Ambiance	"	"

As mentioned earlier, Table 1 is a partial list of varieties that I personally have found to be very productive and more resistant to diseases under conditions where I have worked. Base your choice of variety on the kind of product you want, your indoor conditions and productivity. For example, you may wish a looseleaf lettuce instead of a bibb type, a cherry tomato, not a beefsteak one; you may prefer a certain color of pepper and a particular fragrance of an herb. Your indoor conditions are important. For instance, during the cooler winter months choose varieties that will tolerate lower temperatures and light conditions. Seed catalogs will tell you which varieties do better under lower temperatures and light. Particular varieties will yield higher than others under these environmental conditions, so it is not always the best choice to look only at those highest yielding varieties as they may require much higher temperatures and light than you are able to provide.

Sowing of Seeds in a Medium

You must sow your seeds in some substrate. Some considerations of your choice of substrate include the plant, the hydroponic culture system, water retention, oxygenation, structural integrity, sterility and ease of handling.

There are many different media that can be used to sow your seeds (Fig. 7). A standard method used in raising bedding plants is that of plastic multi-pack trays in flats. The multi-pack trays come in many different compartments per tray. All fit into a standard 10½" by 21" flat. The most common ones for tomatoes and peppers would be the 36-compartment trays. For lettuce and herbs you could use the 72-compartment trays. It is best not to use these trays for cucumbers. With the trays you need a medium such as perlite, vermiculite or a mixture of these with peat. I have found the best is either vermiculite or perlite. These media have better oxygenation than peat.

While this method of using multi-packs is feasible,

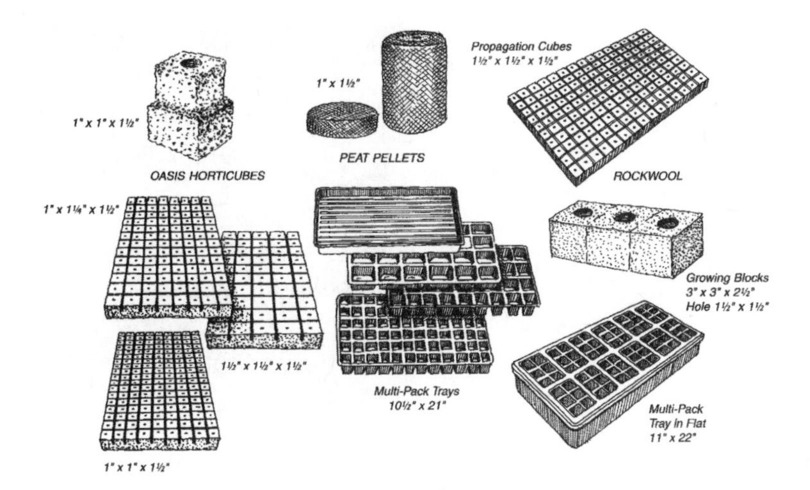

Figure 7. Seedling propagation cubes, blocks, peat-pellets and trays.

it is somewhat inconvenient and messy filling the trays and transplanting. For this reason, most people prefer to use some type of growing cube. You can use Jiffy-7 peat pellets, which are compressed peat contained in a plastic mesh. You soak them in water for 5 to 10 minutes until they swell to about 1" by 1½". You then seed directly into the peat substrate. In some hydroponic systems they may break down and clog your system, so I prefer to use other growing cubes such as Oasis "Horticubes" or rockwool cubes.

Oasis "Horticubes" come as 1-inch by 1-inch by 1½-inch cubes. They are especially good for lettuce and herbs. They provide good oxygenation, retain sufficient water so as not to dry out quickly, are sterile, retain their structure fairly well, have a balanced pH and are easy to handle. They come joined together as 162 cubes that fit into a standard flat. They are easily separated as they are joined only at their bases. I would not use them for tomatoes, peppers or cucumbers as they are very small and are not easily transplanted to a larger block.

The best growing cubes and blocks are those of rockwool (Fig. 8). Rockwool has very good physical proper-

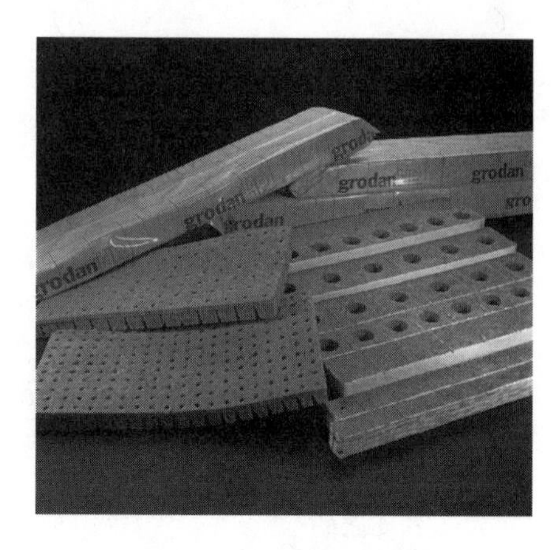

Figure 8. Rockwool cubes, blocks and slabs. *(Courtesy of American Agritech, Tempe, AZ).*

ties of drainage, aeration, water retention, structural integrity, and sterility. Chemically rockwool is inert, but slightly alkaline that can be adjusted by application of water and/or nutrient solution of optimum pH. They are especially well suited to growing tomatoes, cucumbers and peppers. Lettuce and herbs also do very well in rockwool cubes. They are more expensive than the other methods, but for a hobbyist the extra few cents of cost is more than offset by the convenience and success in germination of your seed.

The propagation cubes come in several sizes. The smaller size cubes of 1" x 1" x 1½" high are most suitable for lettuce and herbs. They come in pads of 200 cubes that fit into a standard flat. The larger cubes of 1½" x 1½" x 1½" are best for tomatoes, peppers and cucumbers. Pads of 98 cubes fit into a flat. These cubes have small ¼-inch holes for placing your seeds. Sow your seeds directly into the holes of the cubes.

Place the cube pads in flats and then soak them thoroughly with water. Soak them before sowing the seeds. Be sure that no dry spots are visible after soaking. Use a watering wand on a hose or a hand watering can to soak the cubes. If your seed is fairly new and the viability (percent germination) is 85%–90%, sow one seed per cube with the exception of herbs (not basil) which you can sow 8 to 10 seeds per cube. Basil should be seeded 1 to 2 per cube. Always sow only one seed of tomatoes, peppers or cucumbers per cube. You do not need to cover the seed with any substrate, just water the cubes several times a day with a watering can. This will keep the seed moist during its germination. Be careful when you sow the seed that it falls to the bottom of the hole in the cube. For larger seeds like tomatoes, peppers and cucumbers use the back of a pencil to push the seed down should it not fall all the way to the bottom of the hole.

Tomatoes, peppers and cucumbers (vine crops) are transplanted into rockwool blocks after the seedlings grow to a specific age in the cubes. For tomatoes and peppers use the 3-inch square by 2½-inch blocks.

Figure 9. Tomato seedling in rockwool cube ready to transplant to rockwool block. Seedling is 15 days old. Note that the cube was laid on its side to permit the tomato plant to grow back upwards - allowing roots to grow along base of stem.

Cucumbers are better transplanted to the 4-inch square by 2½-inch blocks. The blocks have round holes 1½" in diameter by 1½" deep to allow the 1½" x 1½" x 1½" cubes to be placed into them. Blocks, like the cubes, must be soaked thoroughly before transplanting. Tomatoes should be transplanted to the blocks at 14 days (Fig. 9), peppers after 21 days and cucumbers in about 7 days. Place the blocks in flats in a checkerboard fashion to permit more spacing among the plants; otherwise, they will get "leggy".

Hydroponic Culture in Choice of Medium

The nature of the hydroponic culture system will also determine what medium is the best to use. For example, when growing lettuce in a floating or raft culture system, you need to use a growing cube that will not fall apart on placing it into the Styrofoam boards.
While Oasis cubes may be used, I find that they crush easily when pushed into the holes of the boards. This causes material to get into the piping of the system and plug tubes. Rockwool cubes are far superior, as they will not break apart as they are pushed into the holes of the boards. They also stay together well when harvesting the plants. A similar problem may arise when using Oasis cubes in an NFT gutter system, however, less so than with the raft system.

Transplanting

Transplanting is the next stage in the growing of plants. In some cases you may seed directly into the hydroponic system, but generally it is better to transplant seedlings into the hydroponic system. This allows you to select the best, healthiest plants. Also, if the seed does not have a high germination percentage, you can sow excess seed to obtain sufficient seedlings to transplant. If you seed direct you must depend on all of them germinating. In a later chapter I discuss seeding direct into pots of vertical plant towers when using a substrate such as perlite. This is applicable to most herbs. It is especially useful if you want to sow many seeds in one pot, as is the case with the majority of herbs.

With lettuce you can transplant the seedlings in their growing cubes rather than going through a second transplanting, as you would do for tomatoes, cucumbers and peppers. Whether you are growing lettuce in Oasis or rockwool cubes, be sure that they are ready for transplanting. Lettuce should have at least 4 to 6 true leaves before transplanting to the hydroponic system. Depending upon their growth rate, it may take up to 3 weeks from sowing the seed to their transplanting stage. For best results with tomatoes, cucumbers and peppers, transplant the seedlings to rockwool blocks. I shall discuss each separately.

As the tomatoes grow in the cubes, they should be separated after about 10 to 14 days when the first set of true leaves have fully unfolded and a second set begins to develop. Here is an important gardening tip with which most people may be unfamiliar. Separate the cubes of tomato seedlings, space them out in a flat laying the cubes on their sides (Fig. 10). If you space them correctly you should get about 4 plants across by

Figure 10. Tomato seedlings spaced in tray and laid on sides.

9 to 10 rows. Leave about 1-inch between rows and cubes. By laying them on their sides the seedlings will bend up and form adventitious roots on the base of the stem. When transplanting to the rockwool blocks either invert the cubes 180 degrees or place them on their sides (inverted 90 degrees). This will cause the seedling to form roots on the stem that will grow down into the rockwool block as the shoot grows up. The result is a very vigorous, healthy plant with many roots. These extra roots, upon transplanting into the final hydroponic system, will reduce the transplant shock and assist the plant in growing very rapidly. Space the blocks in a checkerboard pattern in the trays. About 10 to 14 days later (about 5 weeks from sowing) they will be ready to transplant to the hydroponic system when they have about 3 sets of true leaves. If you want to keep them longer before transplanting, space them again to about half the plants per tray (about 6 plants/tray).

Peppers grow slower than tomatoes, so they will need to be separated as they form the second set of true leaves also, but that will take at least 3 to 4 weeks. Follow the same procedures as with the tomatoes by laying them on their sides at the same spacing in a flat (Fig. 11). After about 2 more weeks they can

Figure 11. Pepper seedlings spaced in tray on their sides.

Figure 12. Pepper seedlings transplanted to rockwool blocks and set in tray in a checkerboard pattern to utilize space efficiently.

be transplanted to the rockwool blocks. You can partially lay them on their sides as you place them in the rockwool blocks. They, similar to tomatoes, will form adventitious roots and should have about 3 sets of true leaves unfolded when ready to transplant as shown in Figure 12.

Cucumbers must be cared for somewhat differently from tomatoes and peppers. Cucumbers grow very rapidly. Their leaves expand quickly and to a large size. For this reason, it is important to separate and space them out early. As soon as the first true leaves have unfolded fully and the next set are visible in the growing point separate the cubes and space them to about half the number as for tomatoes in a flat (3 by 5 = 15 plants/tray). But, do not lay them on their sides, as they are susceptible to such fungus diseases as gummy stem blight that may infect them in a horizontal position. You can in fact skip the step of just separating them in a flat and proceed immediately to transplanting them to the larger 4-inch rockwool blocks. Set them in the blocks upright. Set the blocks in a checkerboard arrangement to get proper spacing (Fig. 13). You can

Figure 13. Cucumber seedlings (17 days old) growing in rock-wool blocks spaced in a checkerboard pattern.

Figure 14. Transplant European cucumbers to bato buckets at 3-week stage. Note the drip lines in the perlite medium and one in the block. Move the drip line from the block to the medium after 7 to 10 days when the plant becomes established in the substrate.

keep them at this spacing for 7 to 10 days. If you do not transplant them to the final hydroponic system, you will need to space them out again to about half the plants per flat (twice the spacing). Do not hold them in the flats more than about 2 to 3 weeks before their final transplant destination (Fig. 14).

Seedling Temperatures

I am aware that it may be difficult to maintain optimum temperatures for your seedlings when growing them in your house, as it is not feasible to keep one room where your plants are located at different temperatures from the rest of the house. The only exception may be in your basement. The second difficulty is that you will be growing all of your plants in one area, therefore you cannot have different temperature regimes for different crops. I will, however, give you this information, so that you can use it as a guideline. Tomatoes require from 77 to 79 F (25–26 C) during germination. As they grow you can lower the day temperature to 73 F (23 C) and the night temperature to 68 F (20 C) for the seedling stage.

Cucumbers like slightly higher temperatures. They take about 2 days to germinate under a day temperature of 79 F (26 C) and night temperature of 70 F (21 C). Upon transplanting to the growing blocks the temperatures may be lowered by about 5 F (3 C) to give a day temperature of 73 F (23 C) and night temperature of 68 F (20 C). That is about the same as for tomatoes.

Germinate peppers between 77 and 79 F (25–26C). As the seedlings emerge lower the temperature to 72–74 F (22–23 C). After transplanting use the same temperature regime as for tomatoes.

Lettuce germinates well under temperatures of 59 to 68 F (15–20 C). Temperatures in excess of 73 F (23 C) may cause seed dormancy. If dormancy is a problem germinate them in a refrigerator at 61 F (16 C). As soon as the seed breaks place them under lights. The best temperature ranges for herbs is from 65 to 75 F (18–24 C). Lettuce seedlings can be grown in temperatures a few degrees lower than for tomatoes, but if you are growing your other seedlings in the same area use the tomato temperature range for best results for all seedlings.

Light for Seedlings

When growing in your home you need to provide artificial lighting for your plants, including the seedlings. As soon as the seeds germinate, that is they break their seed coats, be careful to give them sufficient light or they will get very "leggy". Lighting will be discussed in more detail in the next chapter. Here all I want to point out is that you need to have at least 5500 lux (510 foot candles) intensity at plant surface for 14 to 16 hours per day. You should purchase a light meter to test that you have sufficient lighting at the surface of the plants. However, you must not lower the lights too close to the plants or you will add a lot of heat that may force them into leggy growth.

Strawberries

If you wish to grow strawberries you need to purchase plants, you cannot grow them from seed. Everbearer or "Day Neutral" varieties will produce under similar conditions to those you will be using for other crops in your hobby unit. Some of the more popular varieties include: Seascape, Tribute and Tri-Star. If you purchase actively growing bare-root plants they will produce in about 6 weeks from transplanting. It is best to remove all flowers for the first 3 weeks to allow the plants to get established. Also, remove any runners if they form, otherwise, these vegetative structures will take the nutrients away from the parent plant that needs all its energy to produce fruit. Short-day varieties can be grown over the fall–winter if you keep the daylength under 12 hours. Short-day varieties are normally sold in the fall. Some short-day varieties include Camarosa, Chandler, Oso Grande, Sweet Charlie and Douglas.

Also, consider if you have a backyard garden and are growing some strawberries, you could propagate some runners or lift some plants from the garden in the fall and plant them to your hydroponic unit. Store them in the refrigerator for 3 to 4 weeks to induce dormancy. Stressing the plants in the garden will also cause them to go dormant and thereafter be able to resume growth in your hydroponic unit. Stress them by reducing the watering and fertilizing. Transplant them to your hydroponic unit by mid- to late-September.

3 Cultural Practices

Plant Spacing

Plant spacing is determined by the size of the mature plant. You must consider the actual floor area as the basis for your plant spacing. For example, if you have a small hydroponic unit of dimensions 2 feet by 4 feet, that gives you a total area of 8 square feet. Tomatoes require 3.5 to 4 square feet per plant, but that is floor area. You may think that you can only locate two tomato plants in the hydroponic unit, but not so. The roots of most plants can be contained in fairly small areas as long as they receive adequate oxygen, water and nutrients. In reality you could grow up to 8 tomato plants in that 8 square feet of the hydroponic system. But, you must train the plants outward in a V-cordon method so that at the top of the support cables the total floor area occupied by those 8 plants under your lights would be equivalent to 28 to 32 square feet. That would give the plants adequate spacing so that light could enter the canopy and you would have sufficient access for training the plants.

Peppers require similar spacing although their training is somewhat different as will be discussed shortly. You can plant some peppers and tomatoes together in the same hydroponic system. Perhaps the best would be peppers on one end and the tomatoes on the other, do not mix the plants within the row as that may make training more difficult.

Cucumbers require more area, generally 10 square feet per plant. Their leaves are very large, including the lower ones, so you could not grow as many plants in the system. The maximum would be about 4, which is half the number of tomatoes or peppers. Four plants would need to be trained so that they occupy 40 square feet of floor area.

Lettuce needs to be spaced according to the actual area of the hydroponic system. Lettuce should be spaced either 6" x 6" or 7" x 7" (Fig. 15), depending upon variety. Bibb or buttercrunch varieties can be spaced closer than looseleaf or oakleaf types.

Herbs can be spaced fairly tight. Depending upon their final growth habit they may be spaced 3" x 3" or 4" x 4" according to the area of the hydroponic unit. Also, remember to seed up to 10 seeds per cube or direct seed in clumps. Basil must be spaced similar to

Figure 15. Bibb lettuce at 6" by 6" spacing in a Styrofoam board of a raft culture system.

lettuce as individual plants. Arugula can be seeded directly at 1- to 2-inch centers in rows 6 to 8 inches apart.

Watering

You need to adjust your irrigation cycles as the plants grow. As their leaf area increases and fruit formation is underway the plants will demand more water, so you will need to shorten the irrigation cycles. That is, increase the number of irrigation cycles per day. Fully mature tomato plants may use as much as two quarts of water a day. But, this is also dependent upon temperatures and light intensity. Under indoor growing you should irrigate during mature stage of these vine crops at least once every two hours. Of course, this applies to soilless culture systems that use some substrate as perlite, vermiculite or gravel. NFT and water culture systems have a constant flow 24 hours per day.

Install an automatic float valve system in the reservoir of your unit, so that water may be added to the nutrient tank as the plants use it. This will prevent a possible disaster of running out of solution if you are away for several days. Use a standard float valve assembly similar to your toilet reservoir. You can purchase complete units that are easily installed in any pipe at most irrigation stores.

Temperatures

Optimum temperature ranges differ for each crop. This, however, is not possible for a small indoor unit. It is best to use a temperature regime that is easily achieved in your home as well as being within acceptable ranges for all of your plants. Night and day temperatures should differ by 5 to 10 F (3–5 C) degrees with night temperature being the lower. For most warm-season crops like tomatoes, peppers and cucumbers a temperature range of 60 F (16 C) night and 75 F (24 C) day is suitable. For cool-season crops like lettuce night temperature should be about 55 F (13 C) and during the day (light period) 60 to 65 F (16–18 C).

Herbs can withstand a very broad range of temperatures. They will do best, however, at a range similar to that of tomatoes. Under too low temperatures growth will be slowed, whereas under too high temperatures the plants will get "leggy" resulting in soft succulent plants with little fruit formation. In tomatoes the appearance of purple coloration on the undersides of the lower leaves indicates temperatures are too low.

Keep in mind that your lights will also give off heat, raising the temperature near the plants. Locate lights above the crop sufficiently to get the necessary intensity without adding a lot of heat. You can purchase a thermograph to monitor the temperatures on a 24-hour basis. Each chart is good for one week. These instruments are better than a max–min thermometer having high and low indicators as a thermometer only tells you what was the maximum or minimum. It does not tell you the fluctuation of the temperature over time, as does the thermograph.

Light

Plants have evolved under natural sunlight. To date no ideal artificial lighting can provide the same quality or quantity of light that the plant would like that it receives under natural sunlight. I wish to outline some types of artificial lights that are available for use indoors so that you may better understand what may be the best for your conditions and the plants you wish to grow. As mentioned earlier you want a minimum intensity at leaf surface in the upper portion of the plants of 5500 lux (510 foot candles) for a period of 14 to 16 hours per day.

Types of Lights

In the past fluorescent lighting was the most common form of plant supplementary lighting. Some companies promoted "Gro Lux" lamps specifically for plants. I usually found that cool-white high intensity fluorescents were the best and cheaper than the plant grow lights. Today, however, lighting for plant growth has become more sophisticated. Most lights sold today are of the high-intensity discharge (HID) type. There are two types of lights, high-pressure sodium (HPS) and metal halide (MH). Usually a combination of both gives best results. Also important in your choice of lights are the reflectors.

HPS bulbs are available in 150, 250, 400, 430, 600 and 1000 watts. MH bulbs come in 175, 250, 400, 1000 and 1500 watts. The HPS lights provide more energy in the red part of the spectrum, which promotes blooming and fruiting, while the MH lights are more intense in the blue causing rapid growth. Commercial greenhouse growers will combine MH and HPS lights for seedling growth to be sure that the entire visible light spectrum is available to the plants. This is not practical for indoor growing. Hobbyists have found that Metal Halide light has a wider spectrum and therefore is more useful for indoors where no natural sunlight is present. High Pressure Sodium lights are better to supplement sunlight to extend the day length or to increase intensity during cloudy periods. Metal Halide lamps range in price from $40 to $120 according to their power.

Some manufacturers now make a "conversion lamp". This permits conversion from HPS to Metal Halide. The HPS bulbs are made to operate with MH ballasts. You can interchange the bulbs using the same ballast and fixture. The costs of these lamps range from $75 to almost $200 depending upon their wattage.

Amount of Light

Most vegetable crops including herbs need 30 to 60 watts of light per square foot of growing area. This is a general guideline. Fruiting vegetables such as tomatoes, cucumbers and peppers I have always found like a lot of light, so approach the higher figure for them. To calculate the correct wattage of light needed for a specific area, simply multiply the desired wattage of the light by the area in square feet. Using our earlier example from the "Plant Spacing" section, we had 8 tomato plants that would eventually cover 32 square feet of floor space (8 plants x 4.0 square feet/plant). We need at least 32 x 40 watts = 1280 watts. So you should use a 1500-watt MH light. This 1500-watt MH will provide 47 watts per square foot of growing area.

Lights need to be supported by chains. I have found that using jack chains to support them allows you to change their position as the plants grow. Start the plants immediately with the light. Keep the light at about 3 feet above to get 510 foot-candles at plant surface. Then, lower the light about a foot a week until it is about 1 foot above the plants as they grow upward. Then keep moving the light up several times a week in increments of 6 inches as the plants grow up 6 inches. When the light is close to the plants you must take care not to burn them. Lights give off heat so you must dissipate the heat by using a fan that will move the air across the leaf surface. Remember as I discussed earlier you want to control your day temperatures about 75 F (24 C). You can make more efficient use of the light in the plant canopy by use of a reflector. Most of us just naturally take that for granted. We do not want to have the light going up above the plants, just for it to focus on the plants themselves.

Reflectors & Wall Covering

Reflectors come in three basic forms, parabolic, horizontal and conical (Figs. 16–18). Claims are made that

Figure 16. Parabolic light reflector.
(Courtesy of Light Manufacturing Co., Portland, OR).

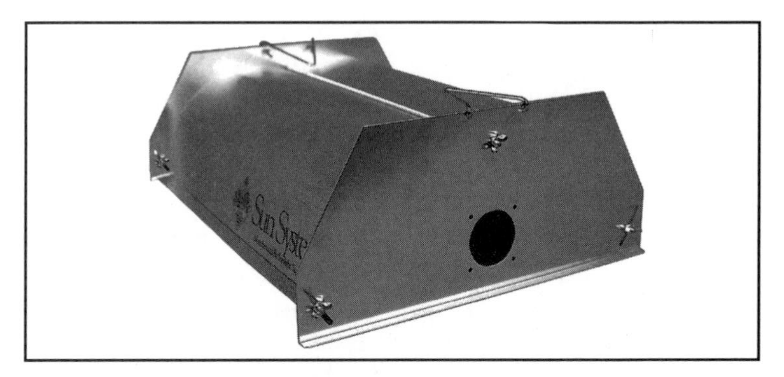

Figure 17. Horizontal light reflector.
(Courtesy of Light Manufacturing Co., Portland, OR).

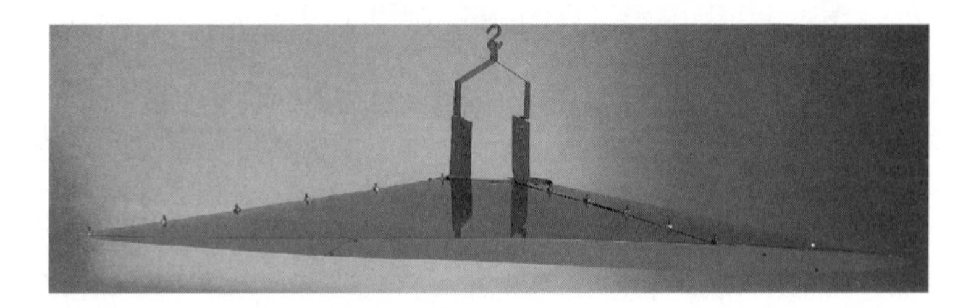

Figure 18. Conical light reflector.
(Courtesy of Light Manufacturing Co., Portland, OR).

parabolic reflectors give up to 18% more light than the conical ones. They are supposed to focus more of the light on the plants by directing the light below the horizontal plane. This also reduces glare to your eyes. Conical reflectors give more side light so they are more useful if you want some light to reach out from the plant canopy. Some reflectors come with a fan to cool the bulb and a glass shield to assist in moving hot air away from the lights. Light Manufacturing Company sells a large "super horizontal lighting system" that uses 1500-watt lamps capable of lighting an area of 80 square feet (Fig. 19). Horizontal reflectors are recommended for HPS systems and where light movers are used. Prices range from $40 to over $200.

Lighting is more efficient when using a reflective Mylar covering on the walls surrounding your hydroponic unit to reflect the light back into the plants from the sides. Mylar will reflect up to 98% of the light striking it. It is available in rolls in 1- or 2-mil thickness.

Figure 19. "Super Plus" horizontal lighting system. *(Courtesy of Light Manufacturing Co., Portland, OR).*

Ballasts

To reduce heat build-up near the location of the lights, install the ballasts some distance away from your growing area. Keep them up off the floor in an area free of splashing water and where they will not fall. Connect your ballast into a timer so that you can set the hours of lighting to come on automatically. It takes up to 30 seconds for the bulb to ignite and up to five minutes for it to become fully bright. During the warm-up period the light may flicker.

When an MH light is turned off it requires 15 to 20 minutes to cool down. Do not re-start it during this cooling period or you will reduce the longevity of the bulb. Metal halide lights should be replaced every year. High-pressure sodium lamps need only several minutes to cool. They should be replaced every few years.

Prices of ballasts are from $200 to almost $400 depending upon the power and whether you want a single or dual fixture (Fig. 20).

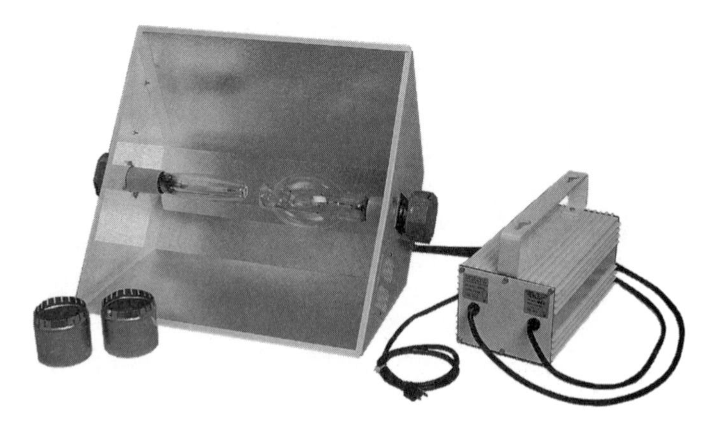

Figure 20. Ballast for an HID light.
(Courtesy of Light Manufacturing Co., Portland, OR).

Light Movers

Light can be more evenly distributed over the plant canopy with some form of moving the lights. Light movers will also help to prevent burning leaves and plants tending to grow toward the lights. A linear mover will move the light back and forth on a track. Tracks may be from 2- to 6-feet long. The lights travel about 2 feet per minute. These are most suitable to a long, narrow growing area. Circular movers are better for more square growing areas. The revolving 360-degree units may support from one to three lights (Fig. 21). A three-arm mover with 3 lights can cover an area 10 feet by 10 feet. They revolve at 16 revolutions per hour. A two-arm, 180-degree rotating system "Sun Twist" (Fig. 22), that holds up to 5 lamps is available from Light Manufacturing Company at just under $280.

Do not be fooled by claims made that light movers allow you to plant more plants at a higher density than we already discussed. The light movers help to distribute the light more effectively over your crop, but they do not increase the energy given to your plants; that is fixed by the wattage and number of lights. However, you can increase growth by reducing the mutual shading among leaves of the plants, which may result in more even growth.

Prices of linear movers are $165 for a 6-ft rail and $200 for a 9-ft rail. A circular two- arm mover costs just under $350 while a three-arm one costs about $450.

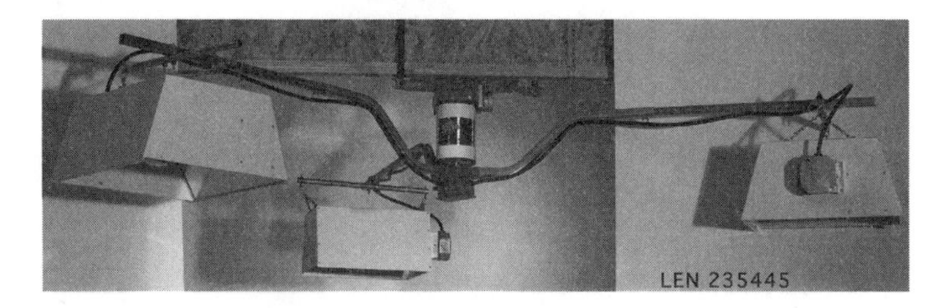

Figure 21. Three-arm circular light mover.
Courtesy of Light Manufacturing Co., Portland, OR)

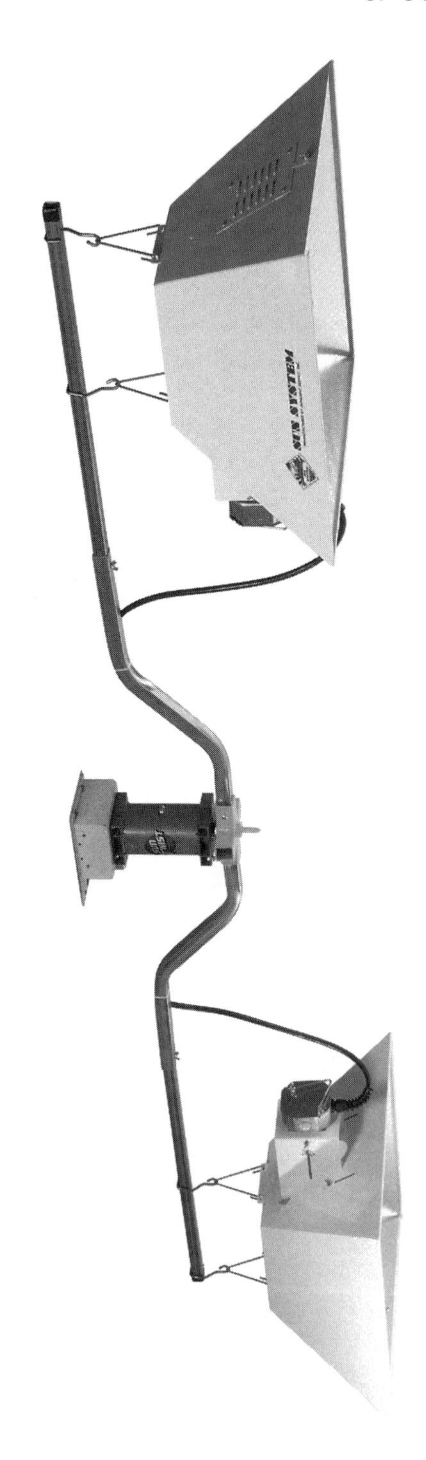

Figure 22. "Sun Twist" two-arm, 180-degree rotating light system. *(Courtesy of Light Manufacturing Co., Portland, OR).*

Carbon Dioxide Enrichment

All commercial growers use carbon dioxide enrichment to increase productivity by up to 20 percent. While such a system is not critical for an indoor grower, it will increase yields. The ambient level of carbon dioxide is about 300 ppm. In your house it probably is lower than that. Research has found that a level between 800 and 1200 ppm is optimum for most plants. There are several types of carbon dioxide enrichment systems available for hobbyists. A system of bottled gas and small delivery tubes to the plants is feasible, but somewhat cumbersome, as the gas tanks have to be refilled. The tanks may be very heavy and awkward to move. A better system is one that generates carbon dioxide by combustion of natural gas (Fig. 23). These small generators, priced from $350 to $550 depending upon their capacity, are hooked up to your natural gas pipe. They have a safety pilot light shutoff just as your furnace has. They must operate with a timer only during the light period. Such generators produce as much heat as a 250-watt halide lamp, so you need to move the air with a fan and/or vent the room to reduce excessive heat. Probably during the light period you would be able to heat the plants from the excess heat of the lights and carbon dioxide generator. During the summer months you would have to vent the heated air out of the room to keep temperatures within the desired range.

The carbon dioxide generator can be coupled with a monitor-controller that monitors the level of carbon dioxide. It will activate your enrichment system whenever the level falls below a preset level and turn it off should that level be exceeded.

We have been discussing what environmental factors are required to produce healthy plants. Now assuming that you have all of these components for your plants under control, you now need to learn to care for the plants themselves.

Figure 23. Carbon dioxide generator that uses natural gas. *(Courtesy of Light Manufacturing Co., Portland, OR and Green Air Products, Gresham, OR).*

Caring For Your Plants

Methods for starting your seedlings and transplanting them were described in Chapter 2 on "Starting Your Plants". Under this section we will look at how to train your plants and some accessories that will facilitate this process.

Tomatoes, peppers and cucumbers must be trained

vertically to best utilize the growing area of your room. We use plastic string to support the plants to either a series of hooks or a support wire above. If you are growing the plants in a spare bedroom, I do not think that you want to suspend wires with hooks along the ceiling. I believe it is more presentable to use decorative hooks (available at garden centers) that are used for hanging plant baskets from your ceiling. Use one hook for each plant. After all, how many plants are you going to grow? Probably no more than 8 to 10 vine crops. One European cucumber plant that produces 2 to 3 fruit per week has always been more than enough for our salads. Several pepper plants and 5 to 6 tomato plants should be enough for all the peppers and tomatoes of an average family. Remember to grow at least one cherry tomato. I have never tasted any better tomato than the cherries!

Stringing

Tomatoes and peppers will have to be lowered as they approach the ceiling. Indoors you will probably carry your plants for about 6 to 7 months from seeding. That is, two crops per year. We will talk about that later when discussing cropping schedules. At this point, however, it is important to decide on how long you wish to carry the crops as that will determine how much extra string you should allow for the lowering process. Generally, I have found that tomatoes over a period of 6 to 7 months will grow to a length of about 11 to 12 feet. So, if you have an 8-ft ceiling, allowing 2 feet for your lights, that gives you about 5 to 6 feet of usable height for the plants. Use "tomahooks", which are special wire hooks for attaching to your support cable or hook above (Fig. 24). Wind 12 to 14 feet of string on the hooks to give you from 10 to 12 feet of string that will support your plants (2 feet is needed from the ceiling to the tops of the plants at their full height). This extra string will permit your lowering of the plants weekly as

Figure 24. "Tomahooks" for supporting plants.

they reach within 2 feet of the ceiling.

There are two types of peppers, the standard "bush" varieties normally grown in the field and the "staking" greenhouse varieties. Your choice depends upon space and how long you wish to grow any one crop. I have found that the greenhouse varieties are more productive and are more controllable in their training. The bush varieties in a hydroponic system still grow about 3 feet tall, so they need some support with strings as do the tomatoes, but without the use of the tomahooks. The greenhouse ones will grow all the way up to your ceiling, similar to tomatoes, but they grow at a slower rate. When cropping them over a 6- to 7-month period, they will reach about 7 to 8 feet in height. For this reason, it is better to use tomahooks with several feet of extra string. The greenhouse pepper varieties also need to be trained to two stems, as we shall explain later, so you need two strings per plant. Space the strings in a V-cordon fashion, training each stem in opposite directions.

Cucumbers are trained somewhat differently. Support them with a string tied to an overhead hook, but without the tomahooks.

A more convenient string support hook is a "Reelenz" reel of string with a hook all in one. They come pre-wound with string making them easier to work with as you simply unreel the string by pulling on it to lower your plants. There is no need to lift the plants and unwind some string from tomahooks to lower the plants.

Begin stringing your plants soon after transplanting to prevent them from falling over. Attach the strings to the stems of the plants using plastic vine clips. Use 3/4-inch ones for tomatoes and peppers and 1-inch clips for cucumbers, which have larger diameter stems.

Training

Tomatoes are trained to a single stem. Support the tomato plant with a vine clip that attaches to the string with a hinge. Locate vine clips about every foot up the plant stem. Be sure to place them directly under a strong leaf so that they will not slip down, as shown in Figure 25. You can also wrap the string around the

Figure 25. Plastic vine clip to support plants by attaching the hinge to the string and the clip underneath a healthy leaf.

stem occasionally to prevent it from sliding down. Always wrap the string in the same direction so you do not forget from one time to the next which way to go around the stem.

Tomatoes must be trained to a single stem, otherwise they become very leafy ("vegetative") and will not produce a lot of fruit. To do that you must remove all of the side shoots ("suckers") that form between each leaf and the stem (Fig. 26). Do this at an early stage when the suckers are about 1-inch long. Delaying this pruning will permit a lot of nutrients flowing into this vegetative growth and as they get larger more shock to the plant and difficulty in removing them occurs. Anybody who has worked with tomatoes in their garden can tell you that they are very hard on your hands by staining and even causing your skin to crack if you do not wash well after handling them. I always use plastic latex surgeon's gloves when working on tomatoes. Unless absolutely necessary avoid cutting the suckers with pruning shears or scissors as these tools will spread diseases more than your hands.

As the tomatoes begin to form fruit, especially for

Figure 26. Side shoot of tomato plant must be removed early when about 1-inch long.

the first 2 to 3 clusters, the fruit load may be very heavy and cause the trusses to break or kink. This will reduce your production as the nutrient flow is interrupted. To avoid this problem use plastic "truss hooks" to support the fruit. There are two types, a "C" truss hook that has two large "C" hooks on one end that is attached to the plant stem and a smaller hook on the other end that is attached to the tomato truss (cluster of fruit). The other one is a "J" truss hook that has several barbed hooks on one end that attaches to the support string and a hook on the other end that goes on the tomato truss (Fig. 27). Usually it is best to have both kinds, as depending upon the length of the truss and tightness of your support string, one may be easier to attach than the other. There is also another type of truss support that is put directly on the truss stem at an early stage to prevent it from kinking (Fig. 28).

Greenhouse varieties of peppers are trained to two stems. As the pepper plant bifurcates at its first flower (crown flower), train these two stems as the principal ones (Fig. 29). Peppers continue to bifurcate at every flower as the stems grow. You must encourage fruit growth by pruning each additional stem to two leaves and flowers. Initially, remove the flowers at the first and second stem layers (bifurcations) to enable the plant to become vigorous. As with your tomatoes remove the suckers often to prevent the plants getting very vegetative. Support the two stems of the peppers with strings and plant clips in a V-cordon way so that the stems are going in opposite directions to permit light to enter the canopy more easily.

European cucumbers are trained somewhat differently from tomatoes and peppers. Cucumbers are supported in a V-cordon configuration whereby plants are alternately supported to one side and then the other as shown in Figure 30. This eliminates leaves overlapping and permits more light to penetrate the plant canopy. They are supported by strings that are directly attached to your ceiling hooks; no extra string is needed. Use the larger plant clips to support the plants under their leaves. Use one every 2 to 3 feet. It is important to wind the string around the vines as they grow. Do this every day as they grow about 6 inches per day under good light conditions. Wind the string around once between each leaf. There is nothing more

Figure 27. "J" truss hook that attaches to support string and tomato truss to prevent the fruit truss from breaking.

Figure 28. A plastic truss support that attaches directly to the fruit cluster.

Figure 29. Pepper plants are trained with two stems.

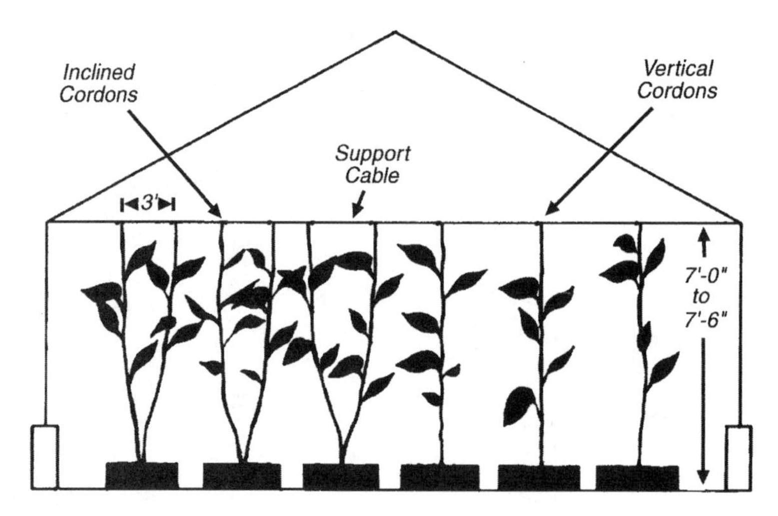

Figure 30. V-cordon system of training.

frustrating than to break a plant! Be very careful as cucumbers break easily. Pull the string down a bit when winding the plant around in a "clockwise" direction. Why a clockwise direction? Well I find most people can remember that. In commercial greenhouses you need to be sure that everybody does it the same or one person will wind them up one day and a different person the next day will unwind them unknowingly and the plants will later fall and possibly break.

When the plants reach the top of the support string you need to pinch the growing point. Allow the two nearest suckers to continue growing. This is the renewal umbrella system of training (Fig. 31). Remove all tendrils as shown in Figure 32 (those monstrous stringing growths that wrap around everything) to prevent them from deforming fruits and leaves. Remove all the suckers that form between the leaves and stem (Fig. 33). Sometimes a second set may form later, so be vigilant to prune all of them as soon as they are about 1-inch long. Pull them off by hand. The first 7 to 8 fruit need also to be taken off so that the plant can become vigorous enabling it to yield heavily later. Remove the small fruits when they are about 1-inch long (Fig. 34). Use these as a special delicacy by deep frying them in

Figure 31. Renewal umbrella system of training European cucumbers.

Figure 32. Cucumber tendrils must be removed.

Figure 33. Cucumber suckers (side shoots) must be removed.

Figure 34. Small cucumber fruit about 1-inch long need to be removed from the plant up to 8 leaf axils to promote initial vegetative growth.

Figure 35. Two laterals trained over the support cable in renewal umbrella training.
(Courtesy of Cuisinart Resort & Spa, Anguilla, B.W.I.)

batter for a few minutes, as you would do with zucchini flowers. They are delicious with the sweet nectar of the flower. The plant will normally only be able to bear 3 to 4 fruit at a time. If the fruit set is more, small fruits above will begin to abort, until the mature ones are harvested.

Train the top two laterals (side shoots) over the support wire or several hooks placed about 2 feet apart and let them grow back down (Fig. 35). Keep removing tendrils. Usually the suckers on the laterals will not grow when fruit is forming. Pinch the growing tip of the laterals as they reach two-thirds of the way down. Once the fruit on the first laterals has matured, cut back the laterals to the next set of suckers near the top of the plant and in this way the growth repeats itself with the next set of laterals. With this method of training you can carry the plants up to 10 to 11 months. However, in my experience with indoor growing, it is better to start new plants after about 4 to 5 months.

Pollination

Cucumbers are all female without any male flowers. For that reason, they do not form seeds. We, in fact, do not want any pollination of the female flowers or seed would form. Tomatoes and peppers need to be pollinated.

In commercial greenhouses we employ bees to pollinate. I do not think that you would want these critters in your house; besides, a small hive has far too many bees for your indoor garden, unless you had a large backyard greenhouse. Pollinate by hand. The simplest method is to use an electric toothbrush (Fig. 36). During the late morning or early afternoon is the best time to pollinate. At that time the relative humidity will be lower and the pollen will flow freely. For 3 to 4 seconds hold the vibrating toothbrush on the truss (cluster) of flowers and watch that the entire cluster moves at high speed. This will release the pollen, which can be seen as a fine yellow powder if you place a piece

Figure 36. Pollinating tomatoes using an electric toothbrush.

of black paper in the background. Pollination is done when the flowers are receptive. This is evident from the flower petals curling back (Fig. 37).

Pollinate at least every other day when the relative humidity is close to 70 percent. If the relative humidity is very high, the pollen will be damp and sticky, so will

Figure 37. Receptive flowers of tomatoes.

Figure 38. Fruit set of tomatoes.

not flow readily. Under very low relative humidity of less than 60% the pollen may dry. Close to midday is generally the best time to pollinate.

You will be able to determine whether or not your pollination is successful within about a week, as small bead-like fruit will start to form (Fig. 38). This is termed fruit set. It is very important to get your plants into a generative state once they start blooming. That is, they will shift their nutrients into more fruit production than just leaf and stem growth (vegetative state). To assist this shift from vegetative to generative state, pollinate every day as the first three flower clusters form. Once these first three clusters have set fruit the plant will become more generative in its growth phase.

Peppers should also be pollinated, but not as vigorously as tomatoes. You must be careful not to break the flowers of the peppers, so do not use an electric toothbrush, just shake the plants by tapping their support strings. Small peppers will appear within a week as they set fruit.

Installing a horizontal airflow fan in the growing room will also assist in pollination. As we discussed earlier, such a fan is needed to move any heat generated by your lights away from the plants to prevent burning and high temperatures within the plant canopy.

Planting Schedules

With indoor hydroponics there is little affect of the season on your growing. Schedule your cropping around the convenience of your needs for fresh product. We want product most during the winter months when high quality vegetables are not available in our local supermarket.

In my experience I have found that it is best to grow lettuce and herbs in a separate system from cucumbers, tomatoes and peppers. This is discussed in more detail in a later chapter on the specific hydroponic units. The main reason for growing the lettuce and herbs separately is that they are very low in form and would not grow well under vine crops when your lights are situated near the ceiling as the vine crops mature. Lettuce also yields greater under a very different nutrient formulation from the vine crops. Finally, the lettuce is on a very short cropping cycle of 30 to 40 days after transplanting, whereas the vine crops continue growing for up to 6 months.

Lettuce can be sown every few days to provide about three transplantings per week. This will give you lettuce every day to harvest. The number of plants to grow depends upon your personal demand for fresh salads.

Herbs, with the exception of basil, grow fairly slowly. They take about 3 to 4 months to get well established. Once they are growing vigorously you can harvest them every day. They will continue to grow for a full year. Basil needs a few months to get established. By keeping it well pruned from the beginning the plants will last up to 4 to 5 months. As soon as the basil reaches 3 to 4 inches tall pinch the growing tip. The plant will then branch out more. Thereafter, keep cutting the tops of each shoot back by 3 inches to another set of small shoots forming between the stem and leaves. Pruning in this fashion will give you a plant that has many branches and will as a result have less woody growth (Fig. 39). That is the secret I have found with growing basil. In fact, I have grown basil for up to 7 months by constantly pruning (harvesting) it. If it begins to flower, it is getting old or is under stress. Pinch all flower buds very early to keep it more vegetative.

Assuming you want tomatoes, peppers and cucumbers most during the winter from November through March, plan the crop accordingly. Tomatoes need to be started no later than mid-August to begin harvesting fruit by mid- to late-November. A seven-month cropping cycle takes them to the end of March. Tomatoes require about 100 days from seeding to first harvest. Begin a second crop by sowing seeds in early March for harvesting to commence in June.

Greenhouse peppers take about a month longer to mature. It would be best to start them in July and again in March. You can use bush varieties that take about a month less to mature than the greenhouse ones.

Figure 39. Pruning basil early results in multi-stemmed plant. Note: place basil in top of plant tower pots.

European cucumbers require two months from seeding to first fruit production. However, if you are going to grow all of these crops together in the same hydroponic unit, start the cucumbers in September so that cucumbers will be ready by November. In this way, the seedlings that are transplanted to your hydroponic system will all start out at the same size and grow up together. This will assist in raising the lights at the correct level above all the plants. It also prevents any older plants from shading out the younger ones.

Overall, then, you will have two crops a year. Start the seedlings in a separate system under their own lights, so as not to interfere with the existing crop as you make the changeover.

Crop Changeover

About one month prior to the scheduled end of your crops, cut off the growing point of your tomato plants. This will shift nutrients to the forming tomatoes. Continue to remove any suckers that develop near the tops of the plants. There is no need to do this for the cucumbers and peppers.

Several days prior to removing the plants, spray them with a mixture of insecticides and fungicide to control the presence of any insects and/or diseases. After this period, one day prior to pulling the plants you can stop the flow of nutrients to wilt the plants down, reducing their weight and volume to facilitate their removal. Compost the old plants if you have an outdoor garden or place them in plastic garbage bags for disposal. Do not keep the string, but you can use the plant clips again if you soak them for a day in a 10% bleach solution. Sweep the floor clean and mop it with a bleach and disinfectant solution. Wipe the walls and ceiling with the same disinfectant. If you are using a Mylar reflective surface on the walls be sure to clean it thoroughly or better yet remove it and replace it with new material.

Sterilize the hydroponic unit with a 10% bleach solution. Run the solution through the system to clean the irrigation lines. Let it sit in the lines for 24 hours. Then, rinse the lines with fresh, clean water several times. If you find some salt formation in the irrigation lines, flush them with an acid and water mixture in the ratio of 1 to 50 respectively. Flush the system thoroughly several times, and then let it sit for 24 hours before rinsing with water.

Change the substrate rather than trying to sterilize it. That is much simpler than trying to sterilize it by baking it in the oven of your stove. Besides, I find that rockwool and even perlite break down in structure during the sterilizing process.

Controlling Pests & Diseases

You may think that it is impossible to get insects and diseases in a well-protected area like a room or basement of your house, but think again, nothing keeps these critters out. Certainly, you can reduce such occurrences by keeping the room isolated from the outside movement of air, but that will not exclude all pests. Keep a close eye on your plants; learn to identify the most common insect pests and diseases. Hang some "bug-scan" cards near the crops (Fig. 40). These are yellow sticky cards that attract the insects. They work like the old standard "fly paper". Once a week identify and count the number of each insect stuck to the cards. Keep this information in a table in a book so that you can refer to it to determine the changes in numbers of insects developing over time. As soon as you discover some pests, you need to take action to reduce or eliminate them or they will get out of hand, feeding on your plants leaving little product for yourself.

Some of the most common pests that you have to identify are whitefly, aphids, two-spotted red-spider mite, thrips, leaf miners, fungus gnats and caterpillars (larvae of butterflies and moths). Whiteflies are by far the most common pest of tomatoes, cucumbers, peppers and even lettuce. To learn to identify them correctly and understand how they develop, refer to some books on pests. Some are given in the references in

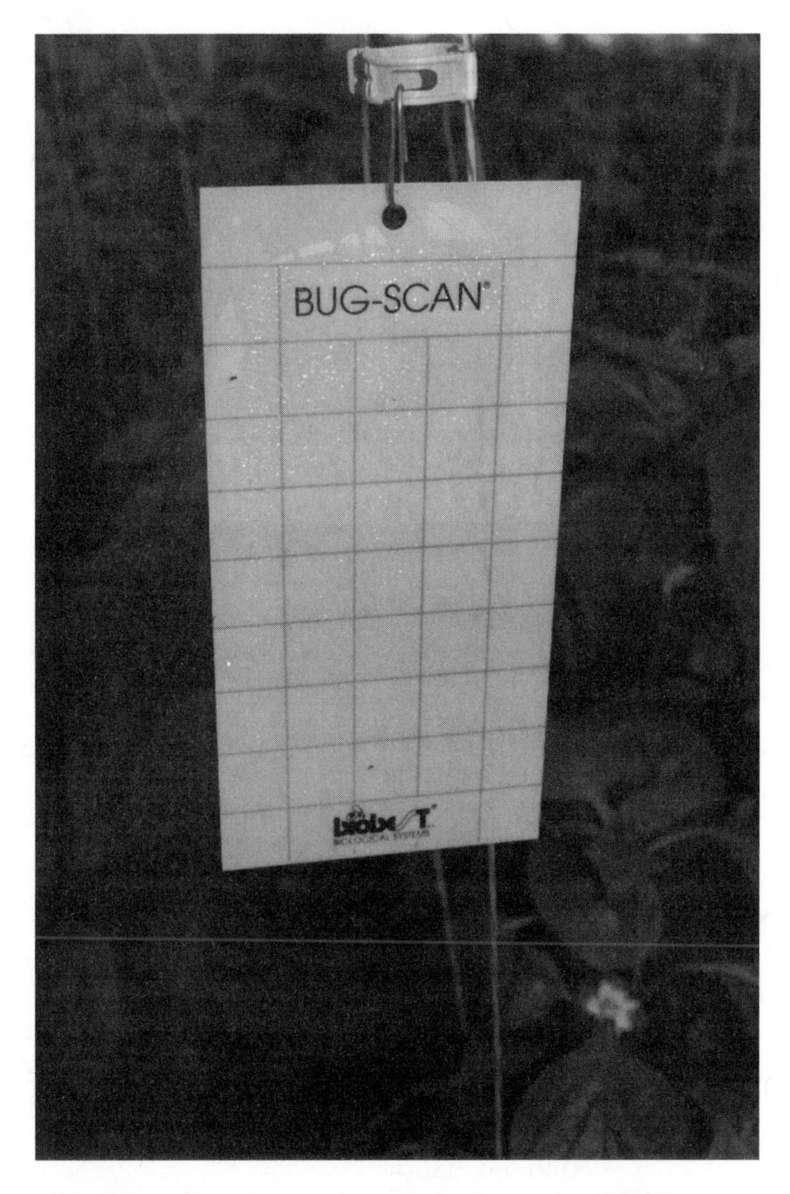

Figure 40. Bug-scan card to monitor insect populations.

Chapter 7. In addition, there are a number of Web sites listed that have photos of these insects. Clear drawings and life cycles are also given in my book "Hydroponic Food Production".

The most difficult part is to determine what control

measures to use on these pests. I hope to simplify that for you here with an outline of some of the safe bioagents that we have used very successfully. Most are available from your local hydroponic shop or from other greenhouse suppliers. Also, you may use "beneficials". These are insects that are predators or parasites on the insect pests. This combined with the use of safe natural bioagents makes up an integrated pest management (IPM) program that is purely biological control.

Briefly I am going to discuss the "beneficials" that help to control a number of the pests. Whiteflies are parasitized by *Encarsia formosa* and *Eretmocerus ermicus,* parasitic wasps. The two-spotted mite can be controlled with the predators *Phytoseiulus persimilis, Metaseiulus occidentalis,* and *Amblyseius californicus. Metaseiulus occidentalis* is more active under cooler temperatures and *Amblyseius californicus* prefers warmer temperatures. Aphids are kept in check with various lady beetles and the green lacewing, *Chrysopa carnea. Aphidoletes aphidimyza,* a midge larva, is also an effective predator. Vertalec, a parasitic fungus, *Verticillium lecanii,* may also be used. Several wasps, *Dacnusa sibirica, Diglyphus isaea* and *Opius pallipes* parasitize leaf miners. *Orius tristicolour* and *Hypoaspis miles* feed on thrips. These beneficial insects come in plastic shake-on bottles or paper strips of pupae that you hang from the leaves of your plants (Figs. 41 & 42). You may purchase these beneficial insects at your local hydroponic shop or from greenhouse suppliers. A very useful reference book on recognizing and determining appropriate beneficials is "Knowing and Recognizing". It is listed in the reference section of Chapter 7.

Other pests may be controlled with various beneficial microorganisms. Caterpillars are very common on lettuce. Control them with "Dipel" or "Xentari", which is a parasitic bacterium, *Bacillus thuringiensis.* Fungus gnats may also be kept in check with *Bacillus* bacterium such as "Gnatrol". An insect-parasitizing nematode, *Steinernema carpocapsae*, controls fungus gnats.

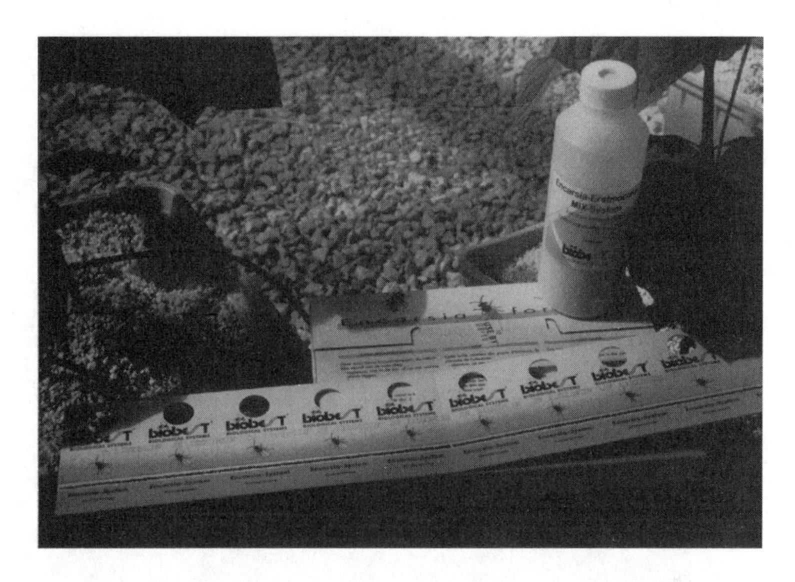

Figure 41. Beneficials available in shake bottles (*Encarsia-Eretmocerus* mix) or cardboard strips of pupae (*Encarsia formosa*).

Extracts from bacteria and plants assist in controlling numerous pests. "Agrimek" and "New Mectin" control mites, "Azatin" or "Neem-X", a plant extract from the neem tree, can be used on whiteflies, armyworms, and mealybugs. "Cinnamite," a cinnamon extract, is good on aphids and mites. Pyrethrins come from the pyrethrum daisy. They are effective contact insecticides against a broad range of pests. "Spintor" controls thrips. "M-Pede", a soapy material from salts of fatty acids, is a contact insecticide effective on aphids, mites, thrips, and whiteflies. Fatty acids are the main component of the fats and oils found in plants and animals. "M-Pede" is also useful as a sticker for applying other pesticides. "BotaniGard" is made of fungal spores that will enter insects and parasitize them. It is used on insects such as whiteflies, aphids and thrips. The addition of one tablespoon of brown sugar per gallon of spray will assist as a sticker for many of these products. It slows runoff of the spray from the leaves and may act as an attractant for some insects such as thrips, which hide in the growing point of the plant and therefore are difficult to contact.

Figure 42. *Encarsia* pupae on strip is hung on a tomato leaf.

Many of these bioagents are compatible with beneficials as they are selective in their action.

Bioagents are also used in the control of diseases. "Phyton-27" is composed of natural minerals and is effective as a fungicide and bactericide. "Cinnamite" offers control of powdery mildew, a fungus disease. Elemental sulfur combined with M-Pede as a sticker is very effective against powdery mildew. “RootShield” and “PlantShield” are nematodes that act as a protectant against root diseases of plants. “Kocide 2000” is a copper compound that helps to control powdery mildew. “AQ 10” is a biofungicide that controls powdery mildew.

I do not want to give you exact rates for usage of these products as they may differ for different makes and situations under which they are applied. Always read the directions and follow them exactly!

Do not be alarmed by this long list of pest problems. Normally, you may be troubled with a few, unless you do not observe your plants and just let things get out of control. Wash your hands and clean your shoes before entering your growing room. Placing a bleach mat at the entrance through which you must walk prior to going into the room will help prevent bringing in organisms from outside. Special mats are available. Keep the mat moistened with a 10% bleach solution.

4 Plant Nutrition

Essential Elements

Plants require 13 essential elements for their growth. In addition to these 13 nutrients they utilize carbon, hydrogen and oxygen, which come from water and the atmosphere. The 13 essential elements are categorized in two groups, those needed in relatively large amounts termed major or macroelements and those consumed in relatively small amounts that are called micro or trace elements. The macroelements include nitrogen (N), phosphorous (P), potassium (K), calcium (Ca), magnesium (Mg) and sulfur (S). The microelements are iron (Fe), manganese (Mn), copper (Cu), boron (B), zinc (Zn), molybdenum (Mo) and chlorine (Cl). Plants cannot live without any one of these elements, hence the term "essential". We as growers must provide all of these 13 nutrients to the plant. In hydroponics they are all added in the nutrient solution.

I often encounter the argument that hydroponic plants are not "organic" because in hydroponics you use fertilizer salts. My reply is that all plants are organic as they only use inorganic ions in their uptake of these elements to grow. The plants manufacture the organic materials of their makeup through photosynthesis. "Organic gardening" is often confused with the fact that in most cases it indicates that only natural bioagents are used to control pests rather than synthetic pesticides. As a result, organic products should really be termed "pesticide free".

Soil in Comparison to Hydroponics

Soil has organic and inorganic components. The organic part is the humus consisting of dead plant and animal matter. The inorganic component is the sand, gravel and rock that must be weathered to release its inorganic elements. The organic material undergoes decomposition by numerous soil organisms and animals. This releases its inorganic elements into the soil water. The released inorganic elements into the soil water give us

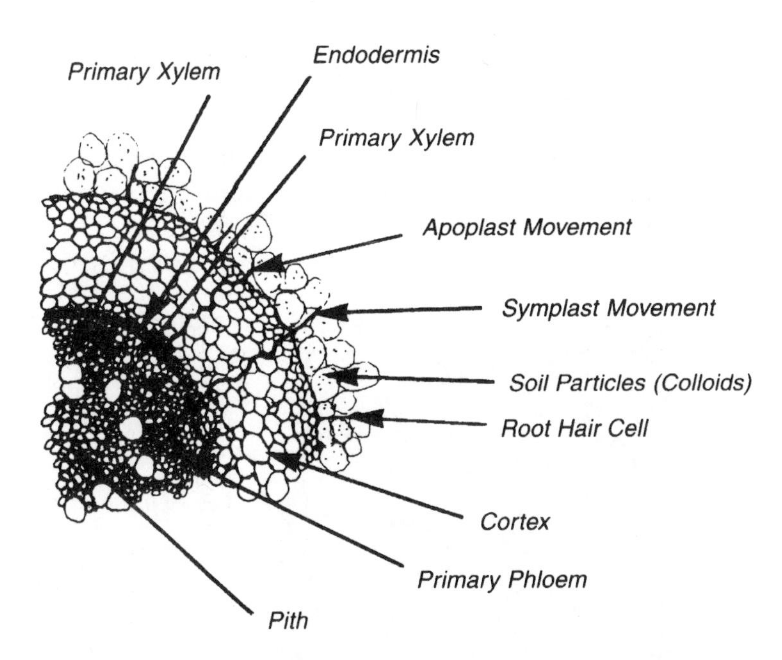

Figure 43. Cross section of root with uptake of water and minerals from the nutrient solution into the vascular system.

the soil solution. This soil solution comes in contact with the plant roots where the individual elements in their atomic ionic state are taken up through a process of electrical and chemical potential difference between the inside and outside of the root membrane (Fig. 43).

In hydroponics we add all of these essential elements from highly pure fertilizer salts. They dissolve in water releasing their individual elements in an atomic ionic state. This is the nutrient solution that contacts the plant roots with differential uptake in the same way as for the soil solution. The difference between hydroponic and soil cultivation is that in hydroponics we add exactly the correct amounts in the optimum ratios of each element so that plant growth is not restricted as it might be in soil when some of these essential elements may be at non-optimum levels. When growing hydroponically, we use bioagents and beneficials as discussed in the previous chapter so our products are pesticide free.

Sources of Essential Elements

In most cases the indoor hydroponic hobbyist will purchase his nutrients prepared from one of the hydroponic outlets (Fig. 44). Generally, the nutrients come in

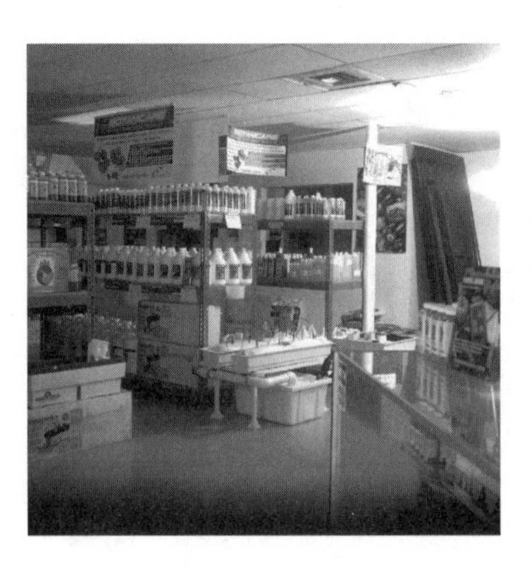

Figure 44. Wide range of nutrients and other hydroponic supplies available at many hydroponic shops. *(Courtesy of American Agritech, Tempe, AZ).*

two components, an "A" and "B" formulation. Two separate components prevent any chemical reaction from taking place in their concentrated form. Usually one part will contain calcium, nitrogen, potassium and iron, while the second one contains the remaining elements. Dissolve them in water separately to prevent any precipitation. The normal ingredients contained in part one (A) include calcium nitrate, potassium nitrate and iron chelate. The second component (B) includes potassium nitrate, potassium sulfate, monopotassium phosphate, magnesium sulfate, manganese sulfate, zinc sulfate, copper sulfate, boric acid and ammonium molybdate.

If you wish to become more involved in the actual nutrition of the plants you may make up your own formulations. However, to do so you need to purchase and store some bags of fertilizers. Do you remember a little about high school chemistry classes that perhaps bored you? Here is your chance to apply it. We will use the chemical symbols for the various salts and units of concentration as parts per million (ppm) or milligrams/liter (mg/l). A general formulation and weights per 10-U.S.-gallon nutrient tank are given in Table 2. You will need a triple-beam balance to weigh these salts in gram units.

Table 2 includes only the macroelements. The weights are very small for a 10-gallon tank, so you must use caution not to make a mistake in weighing the compounds. The weight of the microelements will be even smaller, so we need to make a concentrated stock solution with them and then just add some of that prepared solution to your tank. You can purchase a 10-gallon gasoline tank to store the micronutrient stock solution. Keep it in the dark away from direct heat to prevent algae growth. Table 3 gives you the weights for a micronutrient stock solution of 300 times normal strength in a 10-gallon container.

Table 2. A General Macronutrient Formulation for a 10-U.S.-Gallon Tank

Fertilizer Salt	Weight/ 10-Gal. Tank (gm.)	Elements Supplied and Concentration (ppm)
Calcium Nitrate $Ca(NO_3)_2$	31	Ca: 180 ppm N: 126 ppm
Potassium Nitrate KNO_3	4	K: 39 ppm N: 14 ppm
Potassium Sulfate K_2SO_4	23	K: 250 ppm S: 102 ppm
Magnesium Sulfate $MgSO_4$	19	Mg: 50 ppm S: 66 ppm
Monopotassium Phosphate KH_2PO_4	8.5	P: 50 ppm K: 63 ppm
Totals:		N: 140 ppm P: 50 ppm K: 352 ppm Ca: 180 ppm Mg: 50 ppm S: 168 ppm

Table 3. **A 300-Times Strength Micronutrient Stock Solution in 10-U.S.-gallon Tank**

Compound	Weight (gm)	Element (ppm)
Manganese Sulfate $MnSO_4$	41	Mn: 0.8 ppm
Copper Sulfate $CuSO_4$	3.2	Cu: 0.07 ppm
Zinc Sulfate $ZnSO_4$	5.5	Zn: 0.1 ppm
Boric Acid H_3BO_3	20.5	B: 0.3 ppm
Ammonium Molybdate $(NH_4)_6Mo_7O_{24}$	0.6	Mo: 0.03 ppm

For a 10-gallon nutrient tank you need to add: 10 x 1/300 = 0.0333 U.S. gallons. That is in liters: 0.0333 x 3.785 = 0.1262 liters or 126 milliliters (ml) (1000 ml = 1 liter). Purchase a 100-ml graduated cylinder to measure this volume of solution. If you cannot get the triple-beam scale and graduated cylinder from your local hydroponic shop, contact a scientific laboratory supply company such as Fisher Scientific.

We are still missing iron (Fe), which is also a minor element. Using iron chelate (FeDTPA) that has 10% elemental iron, we need 2 grams in a 10-gallon tank to add 5 ppm of iron.

You now have your nutrient formulation complete with all 13 essential elements.

Water Analysis

Before making up the nutrient solution you should have a water analysis of the local water supply to determine the levels of all of these 13 elements in the raw water. Adjust your formulation for the presence of any of these elements in the water. For example, if the raw water has 10 ppm of calcium, then simply subtract this amount from the original formulation. All of these elements are in ratios, so it is easy to calculate any changes.

Taking our example of 10 ppm of Ca in the raw water, then we want 180 – 10 = 170 ppm of additional calcium using calcium nitrate. We adjust the formulation as follows: 170/180 x 31 gm = 29 gm of calcium nitrate. At the same time the level of nitrogen provided by the reduced calcium nitrate will fall to: 29/31 x 126 ppm = 118 ppm. That is, there are 8 ppm less of nitrogen. That would not influence our optimum level of nitrogen. If there were a larger drop of greater than 10% we would need to add that from another source such as potassium nitrate.

pH of Nutrient Solution

The pH scale measures how acidic or basic a solution is. The scale has a range from 0 (extremely acid) to 14 (extremely basic). These sorts of levels you will not encounter. Most plants prefer a slightly acid condition between 5.8 and 6.5. The pH affects the plants ability to take up its essential elements from the nutrient solution. It also influences the solubility and capacity for the nutrient solution to retain its individual elements in solution.

There are a number of ways to monitor the pH of your nutrient solution. Your choice of pH testing system depends upon the amount of accuracy and reliability in detection you require. While sophisticated pH test meters may be accurate within 1/100 of a pH unit, they really are not practical for other than laboratory use. They also need constant calibration and are very delicate to handle. They will cost from $300 to $500 and upward. Simpler, more durable pH meters are available for just under $100. They too, however, must be calibrated frequently and always kept moist. Other less expensive methods include indicator dye and pH paper. My preference is to use good pH indicator paper that has a range between 4.0 and 7.0. This paper is accurate to within 0.3 units. My favorite is “Merck color pHast® Indicator strips”as shown in Figure 45. These cost about $10 for 100 strips. Paper strips are much easier to handle than liquid dyes that have to be added to a solution sample and then the resultant color of the liquid compared with a color pH chart.

Test the pH of the nutrient solution every day and keep records of it, so you may see any changes taking place. To lower the pH, add acid such as sulfuric acid

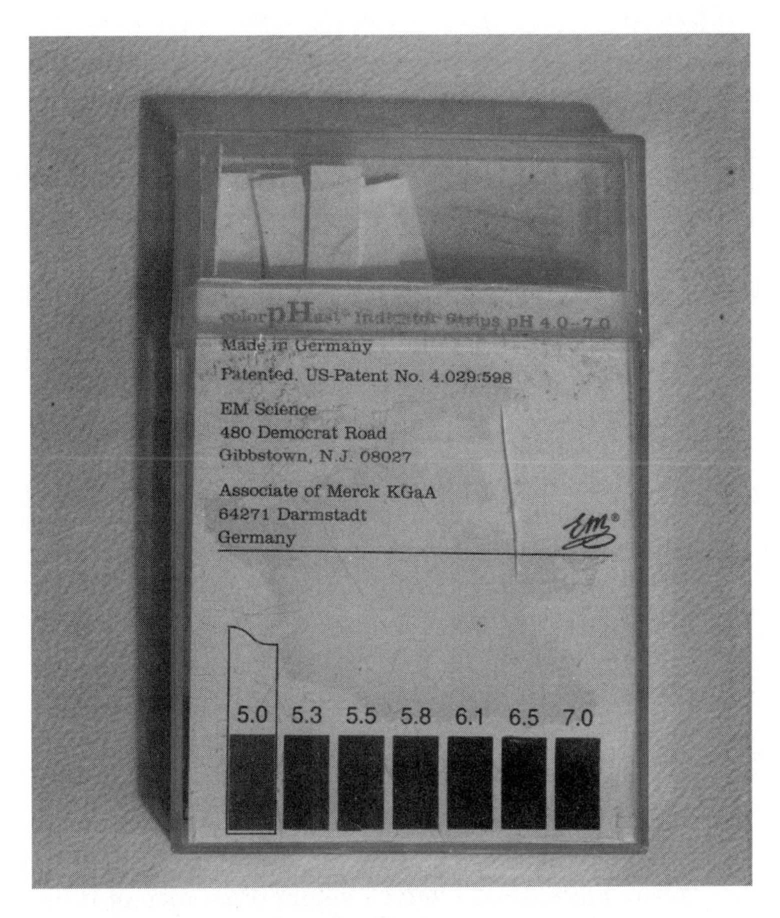

Figure 45. pH indicator paper strips.

(H_2SO_4) (battery acid) and to raise the pH, add potassium hydroxide (KOH), sodium hydroxide (NaOH) or baking soda. You must wear gloves and protective goggles when using sulfuric acid, potassium hydroxide or sodium hydroxide as they burn your skin. These, however, are most effective in quickly shifting the pH. If you buy them in concentrated form, dilute them to about 1 part of acid to 30 parts of water. Remember—always add acid to water!

Your local hydroponic shops will also sell more dilute solutions to adjust the pH that are safer to handle. Do not hesitate to talk to them.

Electrical Conductivity (EC) of the Nutrient Solution

Electrical conductivity is a measure of the nutrient solution's concentration through its ability to conduct electricity. Pure water does not conduct electricity, but any water having solutes (elements) added to it has the capacity to conduct electricity. A special meter, an electrical conductivity (EC) meter (Fig. 46) measures the electricity conducted by the nutrient solution, which is directly related to the level of total dissolved solutes in the solution. The scale is commonly expressed as millimhos (mMhos).

You cannot directly relate parts per million (ppm) to EC as different ions conduct electricity at different rates; therefore, the ratios of these ions as well as their individual concentrations in the solution will influence the electrical conductance. Electrical conductivity does inform you of the overall concentration of all elements within the solution so it can be used as a guide to tell whether the solution has adequate nutrients. But, you could still have an imbalance of individual ions, as the solution may have a high level of a very highly conductive element such as potassium and a low level of one that conducts less electricity such as nitrate ions. So remember, the EC gives you only an idea of the overall total dissolved salts within the solution. To determine

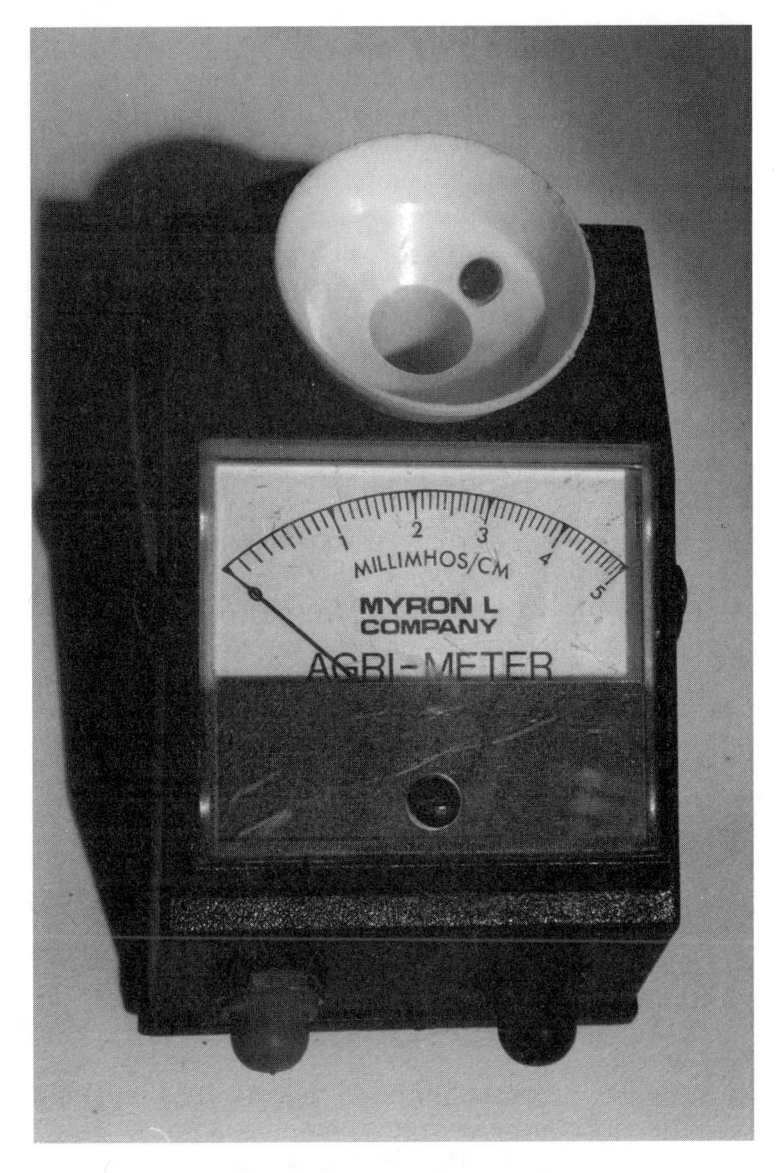

Figure 46. Electrical conductivity (EC) meter.

exact levels of each element, you would have to submit a sample of the nutrient solution for atomic absorption analysis in a private laboratory. Such an analysis would cost about $35 to $50.

Electrical conductivity (EC) meters are essential for

anyone growing hydroponically as they assist in monitoring the nutrient levels of the solution. Generally, a nutrient solution may have an EC of 1.5 to 3.0 mMhos. The best is to test the solution as soon as you make it up. Record this value, then test and record the EC every day. That will tell you of weakening trends in the nutrient solution as the EC goes down over a period of several weeks. At some point, often within a month, the EC will fall enough to justify your changing the entire nutrient solution. The EC meter then indicates at what time you need to change the solution before deficiencies may occur.

Symptoms of Nutritional & Other Problems

If your plants are receiving not enough or too much of any element they will show symptoms of yellowing (chlorosis), browning (necrosis), deformation or stunting of growth. For specific descriptions, functions of elements within the plant and determination of nutritional disorders refer to my book "Hydroponic Food Production". Here I only want to point out some generalities to assist in recognizing nutritional problems with your plants.

Be aware that diseases or insects may also cause symptoms of spots and chlorosis. The first thing to do is to determine whether the symptoms are on the lower or upper part of the plant. The essential elements are grouped as those that are mobile, which can be retranslocated, and those that are immobile, cannot be retranslocated (moved) to another part of the plant. If they are immobile the first symptoms will appear on the upper part of the plant. Mobile elements express their deficiency on the lower part of the plant as they move to the new growth leaving the lower leaves to suffer. Often fruit formation may be affected at the same time. For example, one of the most common symptoms is blossom-end-rot (BER) as shown in Figure 47. This is normally caused by insufficient calcium. It results in tissue breakdown producing a dry, black, leathery appearance of the fruit on the blossom end. The growing points of the plants at the same time will stop expanding and in severe cases will eventually die.

However, environmental factors or watering may also contribute to growth problems. Insufficient watering, too long periods between irrigation cycles, may also cause blossom-end-rot. If plants wilt they will probably form blossom-end-rot of fruit. So, the calcium deficiency in this case is not caused by insufficient calcium in the nutrient solution, but by inadequate watering. Another symptom of insufficient or too frequent irrigation cycles is fruit cracking (Fig. 48). This may occur more frequently under very intense light and/or excess temperatures. Cool temperatures and high relative humidity, especially at the base of the plant, can cause deformation of fruit or what is termed "catfacing" (Fig. 49).

These are only a few examples to make you aware that plant growth disorders may be a result of nutrition, pests, diseases, environmental conditions, or watering. Be vigilant to recognize symptoms and relate to what may be their cause early in the expression of any growth abnormalities. Then, you have time to

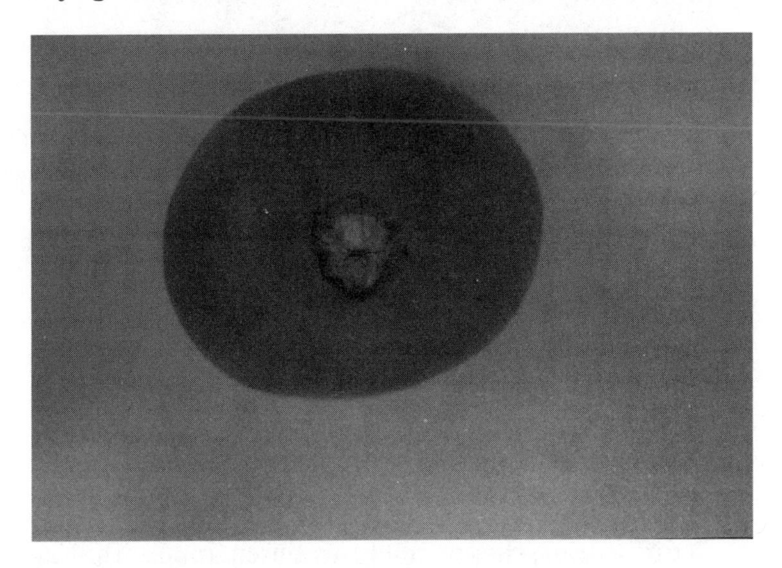

Figure 47. Blossom-end-rot (BER) of tomato fruit.

Figure 48. Fruit cracking of tomato fruit.

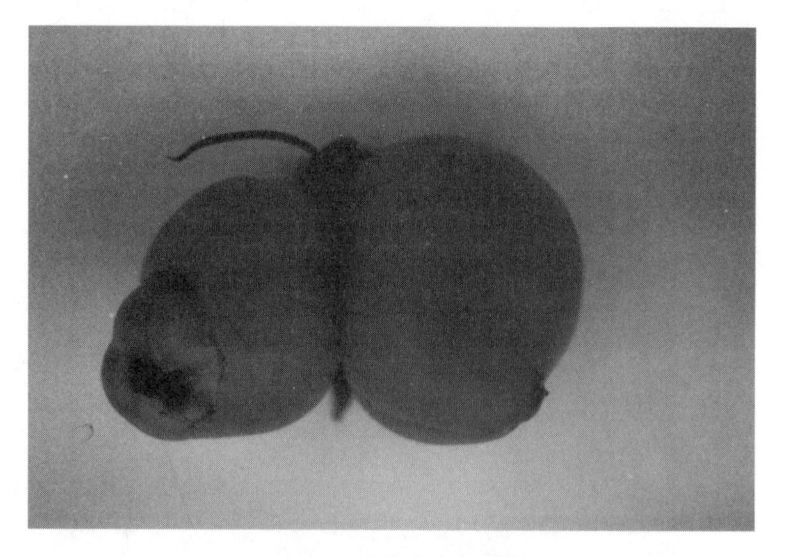

Figure 49. Catfacing of tomato fruit.

correct the problem before the plant becomes very stressed causing reductions in its yields.

While we all hope that we will not encounter such difficulties, these are natural occurrences that keep us learning about caring for our plants.

5 Water Culture (True Hydroponic Systems)

Hydroponics vs. Soilless Culture

By definition "hydroponics" means "hydro" (water) and "ponos" (working), literally "water working". In a broad sense hydroponics means growing plants without the use of soil, but with or without some inert substrate. True hydroponic culture then would be a water culture system without the use of a medium. Today the most common methods of true hydroponic culture are nutrient film technique (NFT) and water culture such as the raft culture system. When a nutrient solution is applied to plant roots from below as a mist this culture is termed "aeroponics". It is also a true hydroponic system.

The most common form of hydroponic culture today is really soilless culture. However, since the same principles of nutrition apply as with true hydroponics most people never differentiate between soilless culture and hydroponics. Many plants, especially longer cropping ones such as tomatoes, peppers and cucumbers, prefer some form of soilless culture over water culture. There are a number of important factors in choosing the best medium for specific plants. The medium must be sterile, that is, free of pests and diseases, have good water retention but not excessive, good oxygenation for plant roots and retain its structural integrity so as not to compact resulting in insufficient oxygen for the roots of plants. Rockwool and perlite are the most popular substrates today. However, depending upon what is available in your area other suitable substrates include peat, coco coir, rice hulls, sawdust, bark chips, vermiculite, sand, gravel, Heydite (porous shale), Leca (clay pellets), and volcanic rock (pumice). Often mixtures of peat, perlite, sand, coco coir and/or rice hulls make a good medium.

Water Culture Systems

At present the nutrient film technique (NFT) is the most popular water culture system. The principle of NFT is to have a shallow flow of water under the roots. The flow must be constant and thin to oxygenate the roots of the plants. If the flow rate is too slow or the film very deep (no longer a film), oxygen demand of the plant roots will not be met and the roots will start to die. Ideally, the upper portion of the root mat that forms in an NFT channel should be above the water level and exposed to air at 100 percent relative humidity.

Both the slope and flow rate of the solution affect the dissolved oxygen content of the nutrient solution. As a root mat forms in the NFT channel a “damming” effect of the root mat can lead to stagnation of the nutrient solution with regions of low dissolved oxygen. To keep the flow of solution from getting stagnant use a slope of at least 1 in 50 (2%) with the NFT channels. It has also been found that flow rates of 2 to 3 liters per minute or slightly greater increase oxygenation. The height of free fall of the solution back to the nutrient reservoir is important in aeration of the solution. More height causes a finer break up of the solution and introduces air bubbles into the tank.

You may also use an air pump with air stones in your nutrient tank to enhance the dissolved oxygen level of the solution. These factors are important in the design or choice of an NFT system whether you build it yourself or purchase a complete ready-to-go system.

Choosing the System

The correct choice of system will improve your success in growing. Ask yourself some of these questions. What crops do I want to grow? How many plants of each crop do I wish to grow? What available floor area do I wish to use? Remember the spacing needs of plants discussed earlier in Chapter 3. You must base the number of plants on the total floor area you have available. Do you want a large system or a series of smaller ones? As pointed out earlier do not grow lettuce together with vine crops, as it will get insufficient light. Keep low profile plants in separate systems from the vine crops. Do you want to utilize your hydroponic units most effectively by starting your seedlings in another system of trays with rockwool cubes and blocks? This gives you a head start on the growing cycle by transplanting older plants.

To help you answer some of these questions consider this. Determine how much salad crops you need for your family. How many head of lettuce a week, how many pounds of cucumbers, tomatoes, and peppers do you consume weekly? Once you have these numbers and types of crops you eat weekly you can calculate the total number needed of each. Remember that you need to sow lettuce at least 2 to 3 times a week and the time for them to mature may take up to 40 days in the hydroponic system.

Lettuce will grow very well in an NFT system, so that should be an obvious choice for it. However, the vine crops would do better in a soilless system. As a result, your choice of the system for your lettuce will differ from that of the vine crops.

NFT Systems

There are a number of manufacturers of NFT systems. My purpose here is to show you some of the units available and discuss their differences. While price ranges are given, these will fluctuate with the different manufacturers and models. American Agritech has an NFT unit they call "Jetfilm®" (Fig. 50). Various models contain channels (NFT gullies) in 4-ft., 6-ft., 8-ft., 12-ft., and 16-ft. lengths. These channels are available in widths of 4-, 6- and 9-inches. Plant spacing within the channels is 8 inches on the 4- and 6-inch widths, and 12 inches for the 9-inch width. The NFT channels in which the plants are grown support net pots of 2", 4" and 6". The pots contain expanded clay pebbles as a substrate to sow seed or transplant seedlings. While I agree that the narrow trays that use the small 2-inch diameter net pots are an NFT system, the larger pots and channels would really be a combination of NFT and rock culture.

Lettuce, herbs and strawberries would be most adaptable to the narrow channels and small pots. The larger pots in the wider channels would be better for vine crops such as tomatoes, peppers and cucumbers. For these use the 9-inch wide channels with 6-inch diameter pots spaced at 12 inches within the channels. The least expensive system of 3 channels (trays) costs under $400 whereas, a large system of 36 trays is about $2300 for the 4-inch wide trays. The costs of 6- and 9-

Figure 50. American Agritech NFT unit "Jetfilm". *(Courtesy of American Agritech, Tempe, AZ).*

Figure 51. "Terrace Hydrogarden" by American Agritech. *(Courtesy of American Agritech, Tempe, AZ).*

inch wide trays range from $400 to $2500 for 2 trays and 24 trays respectively. The systems consist of the trays, support stand, piping, nutrient tank and pump.

They also sell a "Terrace Hydrogarden", which is a tier system in a staircase design in both lean-to and "A"-frame configurations. These are aluminum bench supporting structures (Fig. 51).

American Hydroponics manufactures five models of NFT systems. The regular "NFT Gully Kits" are of dimensions 4 feet by 4 feet or 4 feet by 6 feet and are 21 inches tall (Fig. 52). The NFT channels for lettuce are

Figure 52. "NFT Gully Kit" growing different lettuces.
(Courtesy of American Hydroponics, Arcata, CA).

2" high by 4" wide with 1-inch holes to fit five oasis or rockwool cubes (Fig. 53). The bottoms of the channels have a number of ridges to make the flow of nutrient solution spread across the channel. This prevents dry spots where plants may not get any solution. The kit includes 5 gullies, plumbing, galvanized steel table frame, submersible pump and 30-gallon nutrient reservoir with cover. The 5 gullies have holes for 25 plants of lettuce or basil or clumps of herbs. This spacing of approximately 7- to 8-inches between plants is optimum for these plants. The smaller unit has a price of about $400 and the larger one costs about $430.

Their "NFT Rockwool Gully Kit" comes in the same table dimensions as the one above. The difference is that the NFT channels are 3" high by 6" wide and therefore better adapted to growing vine crops like tomatoes, peppers and cucumbers. The gullies have larger 3-inch square holes to hold five 3-inch rockwool blocks (Fig. 54). This kit comes complete as described

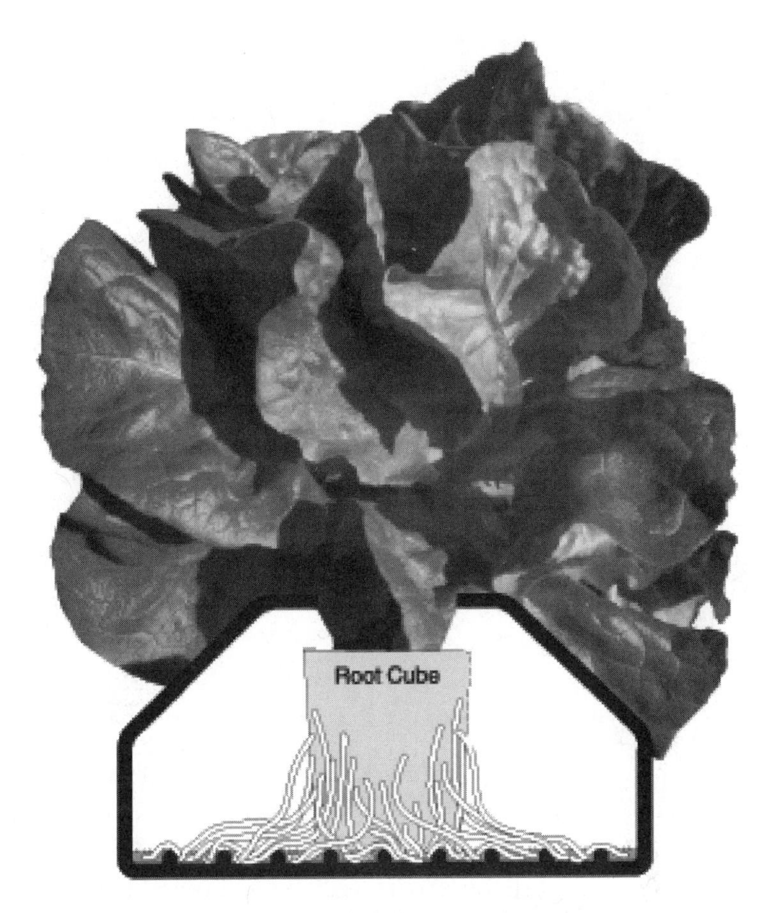

Figure 53. Cross section of NFT channel.
(Courtesy of American Hydroponics, Arcata, CA).

for the "NFT Kit" for a price from $480 to $530 for the small and large units respectively. The 4' x 4' Kit has 23 holes for plants and the 4' x 6' one has 33 holes. From previous discussion on spacing we know that this number of vine crops could not be grown so close together. The area of the table is only 4' x 4' = 16 square feet for the small one and 24 square feet for the larger one. The maximum number of tomatoes or peppers we could grow in this area would be 5 or 7 respectively. About 3 cucumbers could be contained in that area. You could grow more by spacing the tops of the plants wide apart in espalier form, but still 23 or 33 plants would not fit.

Figure 54. "NFT Rockwool Gully Kit" by American Hydroponics.
(Courtesy of American Hydroponics, Arcata, CA).

The other model "NFT Combo Gully Kit" would be a better choice to grow lettuce and herbs on one side in the smaller lettuce channels and some vine crops in the larger channels. This combination of two "NFT Rockwool" channels and three "NFT" channels is more practical, but be aware that you must have one set of lights for the vine crops and another lower down for the lettuce and herbs. There are 9 sites for plants in the two "NFT Rockwool" trays, so you could use every other one or separate the plants well apart on support strings. This combination kit has a price range from $430 for the small model to just under $500 for the larger one.

American Hydroponics has a "NFT Wall Garden" which is like a lean-to arrangement of channels. This is specifically for lettuce and herbs. They claim that it may be mounted on fences, walls, garages or any vertical surface where there is good light as shown in Figures 55 and 56. This would, of course, be for summer use in your backyard. This unit uses the standard 2" x 4" gullies. The complete kit includes three gullies with

Figure 55. "NFT Wall Garden" by American Hydroponics growing various types of lettuce.
(Courtesy of American Hydroponics, Arcata, CA).

Figure 56. "NFT Wall Garden" growing strawberries.
(Courtesy of American Hydroponics, Arcata, CA).

Figure 57. "612 NFT Production Unit" measures 6 ft by 12 ft with 9 production channels and one seedling channel. It is ideal for lettuce and herbs.
(Courtesy of American Hydroponics, Arcata, CA).

24 plant holes, wall brackets, plumbing, submersible pump and reservoir with cover for a price of $270.

The Company also has a "612 NFT Production Unit" that is designed for a small commercial grower or a hobby grower with a large appetite for lettuce and herbs. This unit measures 6' x 12' on a bench 34"–38" high (Fig. 57). It has 9 gullies with 1 3/4" holes at 8-inch centers for the production trays and one gully with holes at 2 inches to act as a seedling tray. The 72 plant sites for the seedling gully permits you to rotate your crops. The production gullies house 162 plants at 8-inch centers. For a price of approximately $1200 the kit has the 10 channels, the plumbing, galvanized steel table frame, submersible pump and a 35-gallon reservoir with cover.

A-Frame NFT

Future Farms produces an A-Frame NFT system that they call Future Farms 196 (FF 196) as it grows 196 lettuce or herbs. The A-frame creates 24 square feet of growing area within 12 square feet of floor space. Plants are started in rockwool cubes and the seedlings transplanted into the A-frame. Plant sites are located at centers of 4 1/2" x 3" so if you wish to put larger plants like basil in the system place them in every other hole and close the unused holes with the special plastic plugs provided with the kit. I grow 196 head of leaf lettuces such as Red Sails, Green Salad Bowl and Red Salad Bowl in this system or a combination of basil, lettuce and bok choy (Fig. 58). However, the spacing is very tight, so harvest the lettuce early, about 3 weeks after transplanting. If you want larger heads use every other hole to get sufficient light for the plants to prevent bolting.

The reservoir, which acts as the base of the A-frame, measures 3 ft x 4 ft x 6" deep. It holds up to 45 gallons of nutrient solution. The sides making up the "A" shape consists of corrugated panels covered by a flat panel

Figure 58. "Future Farms 196" grows 196 plants. Shown here is a combination of basil, bok choy and bibb lettuce on the left side and Red Sails and Green Salad Bowl leafy lettuces on the right side of the A-frame.

Figure 59. Corrugated plastic panels make up the sides of the A-frame where the nutrient solution flows back into the reservoir below.

having the holes for the plants (Fig. 59). A perforated pipe header located at the peak of the A-frame and between the two panels conducts the nutrient solution to each gully in the corrugated panels. The solution runs constantly down the panels and back into the reservoir where it is pumped back. The holes in the growing panels line up with the gullies of the corrugated panel where the nutrient solution rapidly flows down in a thin film aerating as it falls into the reservoir below. The price of this unit is about $350. I have been very satisfied with the performance of this unit and our guests are always very impressed.

Ebb & Flow Water Culture Systems

An ebb-and-flow system floods the plant roots from below and then drains back to a nutrient reservoir (Fig. 60). This is not a continuous flow system as is the NFT, but operates on 4 to 8 cycles per day depending upon environmental conditions and plant growth. A submersible pump in a reservoir below periodically floods a bedding tray aerating the plant roots as the solution enters and drains back to the reservoir (Fig. 61). The plants sit in a tray insert having holes in it to hold either rockwool blocks or growing pots where the plants are seeded or transplanted into a perlite or gravel medium. If rockwool blocks are placed in the plant sites this system is water culture; however, if pots with a substrate are used it is really soilless culture.

American Hydroponics makes several models, "Heavy Duty (HD) One & Two-Tray System" are priced at $370 and $565 respectively. An aluminum table supports one or two 3 ft by 3 ft black ABS plastic trays. A 30-gallon reservoir with cover sits below on a shelf of the frame. A submersible pump in the reservoir pumps up the nutrient solution through a plastic pipe to the flood (bedding) tray. A timer operates the pump. A 3 ft x 3 ft one-tray "Vegi-Table" unit costing $500 uses rockwool cubes (Fig. 60). The tray

Figure 60. The American Hydroponics "Vegi-Table" is an ebb-and-flow system that floods the plant roots from below and then drains back to a nutrient reservoir.
(Courtesy of American Hydroponics, Arcata, CA).

Figure 61. Healthy roots of plants growing in the ebb-and-flow tray of the "Vegi-Table". Note the fill/drain pipe in the base of the tray on the right.
(Courtesy of American Hydroponics, Arcata, CA).

insert for it has 36 square plant sites (1 1/2") in which the 1-inch rockwool cubes fit. This spacing is adequate for lettuce and herbs.

Aeroponic Systems:

Although I personally have not grown in this type of system, the concept is very interesting. A company in Canada, North American Greenhouse Supplies, which markets as "Future Harvest Development", has developed a unique Aeroponic system they named "Pipe Dreams". The grow tubes are 6" (15 cm) in diameter and available in 4 ft., 8 ft. and 10 ft. lengths. The grow tubes are mounted in a tier system on lean-to support frames. This gives them a "V" shape to permit light to reach the lower tiers. Mesh pots fit in plant-site holes of the grow tubes. The mesh pots filled with expanded clay pellets suspended inside the grow pipes receive nutrients in a mist form by a fine spray system inside the grow tubes. Each 360-degree spray head mister, which feeds two pots, is connected to a feeder line attached to 1/2" black vinyl tubing from the pump in the nutrient tank. A timer activates the pump producing a water cycle of 1 minute on and 5 minutes off. On the larger units a 3/4" flexible hose connects to an end cap of each growing tube, which is attached to a 2-inch PVC manifold that conducts the solution to the reservoir below. On smaller units the flexible hose connects directly to the reservoir. A float valve in the nutrient tank allows you to connect the existing nutrient tank to a larger reservoir.

They have eight different sizes of systems. Their basic unit is "Pipe Dreams Balcony 18" which measures 48" x 26 1/2" x 44 1/2" high (Fig. 62). It has 18 plant sites and 9 mist heads. It comes with a 13-US-gallon nutrient tank and 3 grow tubes (6" by 40") supported by a lean-to frame on casters so that it can be easily moved. The suggested price is $567.

The "Pipe Dreams Balcony 32" is the same dimensions as the "18", but has 4 grow tubes (6" by 48") with 32 plant sites. It is also on casters and includes the 13-US-gallon tank (Fig. 63). It costs $620.

Two new models, "V32" and "V36" have a V-frame supporting four 6-inch diameter pipes of 48" or 60" respectively. The smaller unit has 32 plant sites and 16 misters whereas the larger one has 36 plant sites with 18 misters.

The next size is "Pipe Dreams–PD64" which measures 96" x 44" x 38" high that supports 4 pipes 8 feet long having 16 plant sites per grow tube for a total of 64. This model is also a lean-to. "Pipe Dreams–PD96" is two lean-to configurations placed together to form a "V" frame holding 6 pipes (3 per side) (Fig. 64). This holds 96 plants and has 48 mist heads (8 per grow tube). It also has a larger nutrient tank of 55 gallons. The price of this unit is about $1400. The "PD160" is of the same

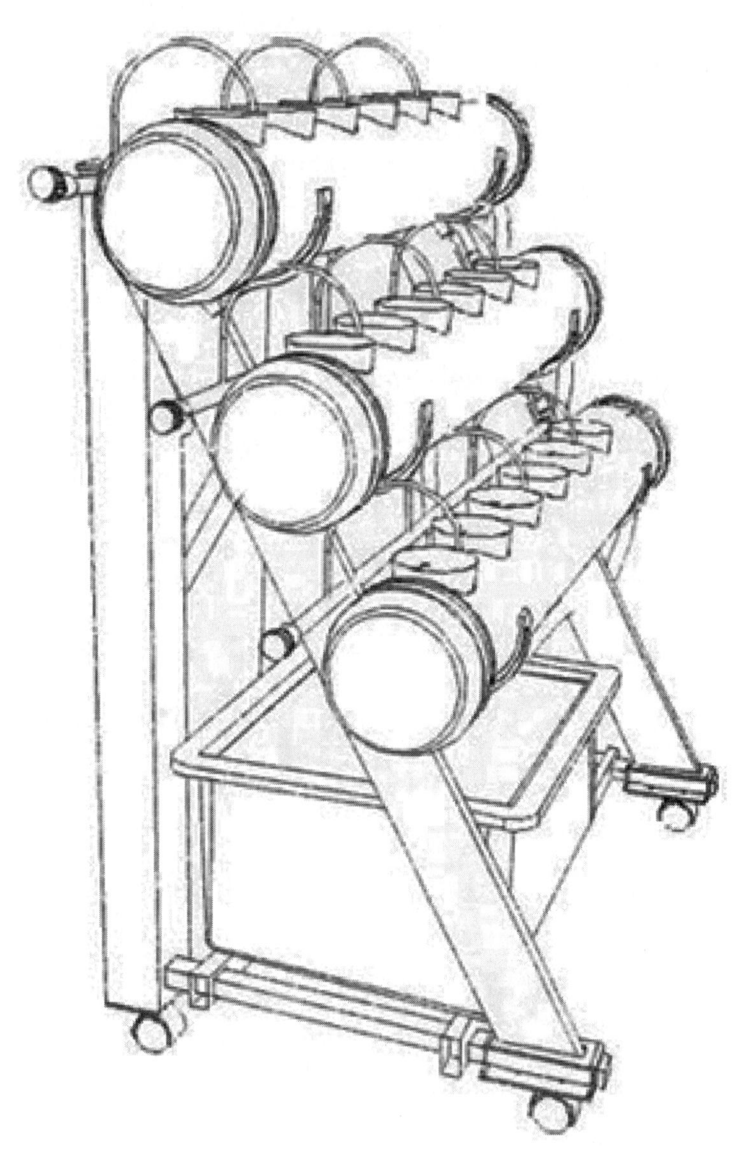

Figure 62. "Pipe Dreams Balcony 18" aeroponic system. *(Courtesy of North American Greenhouse Supplies, Calgary, Alberta, Canada).*

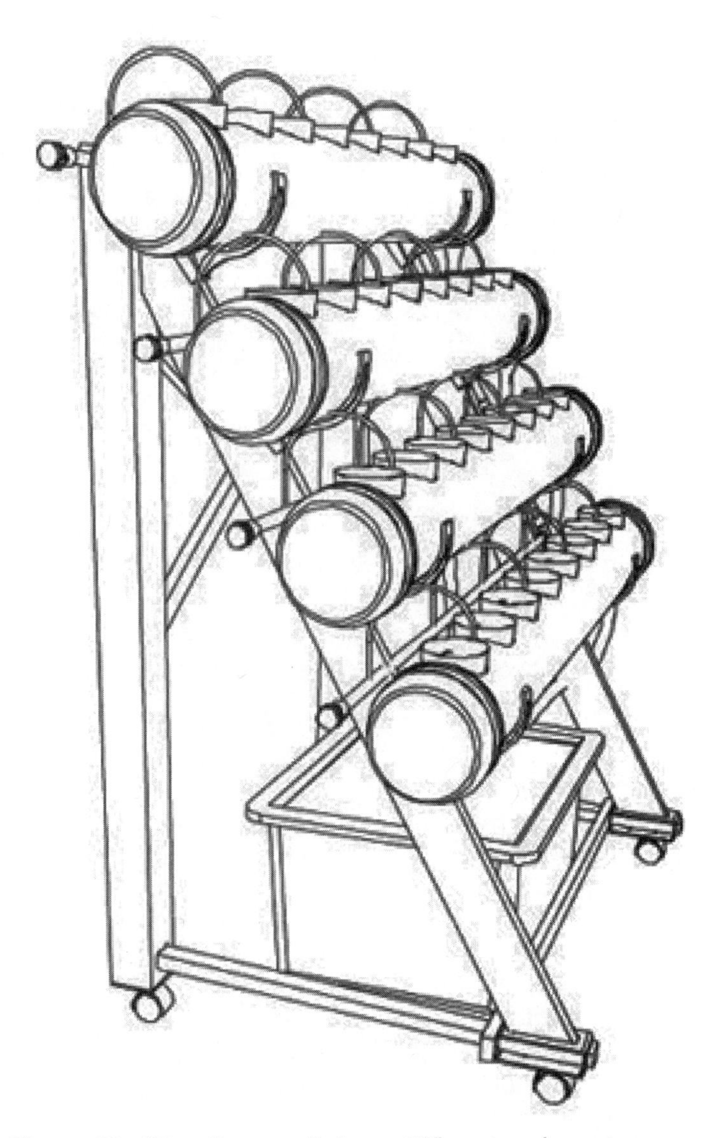

Figure 63. "Pipe Dreams Balcony 32" aeroponic system. *(Courtesy of North American Greenhouse Supplies, Calgary, Alberta, Canada).*

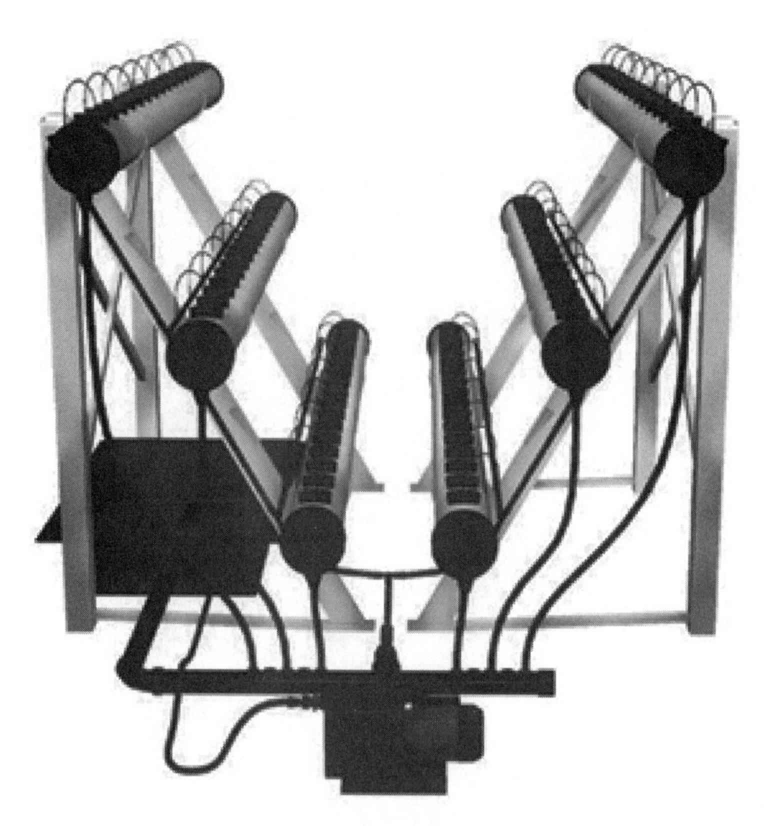

Figure 64. "Pipe Dreams–PD96" aeroponic system. *(Courtesy of North American Greenhouse Supplies, Calgary, Alberta, Canada).*

dimensions as the "PD96" which is 96" x 76" x 58" high. Ten pipes 8 ft long arranged in the "V" frame will hold 160 plants (Fig. 65). The price is about $1700. Their largest unit holds 200 plants and has a 97-gallon tank at a price of approximately $2000. It has ten pipes 10 ft. long.

I think that this system is good for growing lettuce and herbs, but due to the close spacing of the plant sites (about 6 inches apart) vine crops would not have sufficient area. The smaller lean-to units may be suitable for vine crops in the bottom pipe if every other hole was planted and they were trained to slope away from the grow tubes to overhead supports. Remember the principle of 3.5 to 4 square feet of floor area per toma-

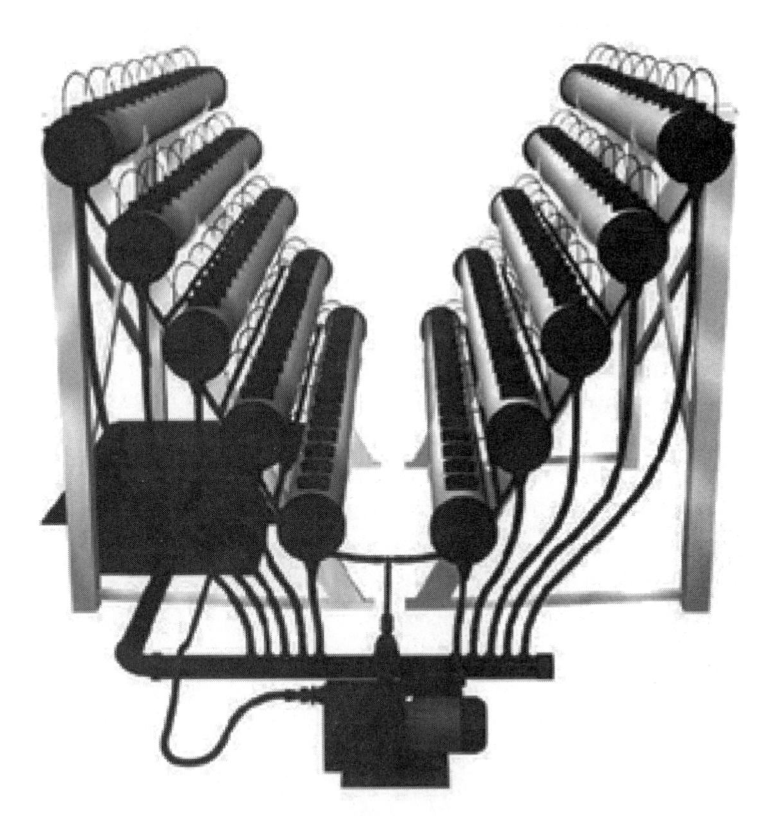

Figure 65. "Pipe Dreams–PD160" aeroponic system is arranged in a "V" frame holds 160 plants.
(Courtesy of North American Greenhouse Supplies, Calgary, Alberta, Canada).

to or pepper plant and 10 square feet for cucumbers. The upper growing tubes would be fine for lettuce and herbs, as they would not shade the vine crops. But, be careful with the lights above. You must have two sets of lights; one for the vine crops and another for the lettuce and herbs as the light for the vine crops would be too high for the other low profile crops. Another possible arrangement of the growing tubes would be to place two lean-to frames back-to-back to get an A-frame configuration rather than a "V" form. This should permit light to enter the vine crops in the bottom grow tubes with the lettuce and herbs above, but you must V-cordon train the vine crops away from the growing system.

Combination Water Culture & Soilless Systems

As discussed above if an NFT or ebb & flow system uses large mesh pots with a substrate such as perlite or pebbles, it really is not a true water culture system, but a combination of water and soilless cultures. If the mesh pots are only 2" in diameter, the system could be termed water culture, as the majority of the plant roots would be in the solution below. But, with larger pots more roots should remain in the substrate and therefore in my opinion it is a combination of soilless and water cultures. This does not mean that the system is less functional; in fact, it may be more suitable to a wider range of plants, especially for vine crops. Vine crops would do best in the larger pots of 5 1/2" or 8" diameter.

Small units of this combination system are available from several companies such as American Agritech, Diamond Lights, Homegrown Hydroponics and General Hydroponics.

The "Hobby Farm" of Diamond Lights is made of polystyrene plastic and measures 21 1/2" x 21 1/2" x 10 1/2" high. It features a 7-gallon reservoir beneath a removable plant tray that contains 9 plastic pots with a lightweight rock medium (Fig. 66). The kit includes pump, hose and fittings, growing medium and nutrients. Homegrown Hydroponics has several similar ebb-and-flow units that will grow 12, 17 or 31 plants. They also sell two sizes (2 ft. and 4 ft.) of mini NFT units that hold 4 or 8 plants. Prices start at under $100 for the smallest NFT unit and range to $320 for the largest ebb-and-flow garden. These units will grow lettuce, herbs and small patio tomatoes.

Another small ebb-and–flow unit of this nature is the "Baby Bloomer" by American Hydroponics with dimensions of 31" by 12" by 12" at a price of about $170. This a self-contained unit with a reservoir underneath supporting a bedding tray where the solution is pumped

Figure 66. The "Hobby Farm" of Diamond Lights is made of polystyrene plastic.
(Courtesy of Diamond Lights, Petaluma, CA).

periodically to flood the bases of the mesh pots containing the plant roots (Fig. 67). A tray that sits on top of the bedding tray supports ten mesh pots having perlite or rocks as a medium. While this unit is fairly small it would grow a combination of lettuce, herbs and perhaps several small "patio" type cherry tomato plants.

They have several larger units called "One & Two-Tray Econo System", which are the same as was described in the above section on ebb-and-flow. The difference is that these units use 23 mesh grow pots with an expanded clay medium per 3 ft by 3 ft growing tray instead of the 36 sites for rockwool cubes. At that spacing you could grow lettuce, herbs and a few vine crops providing you train them correctly as described earlier. Prices are comparable to those discussed under "ebb & flow".

A small ebb-and-flow system by General Hydroponics has six channels sitting on top of a nutrient reservoir (Fig. 68). It holds up to 42 plants as lettuce, arugula, spinach, etc. (Fig. 69).

General Hydroponics presents a somewhat differ-

Figure 67. "Baby Bloomer" ebb-and-flow unit growing dwarf patio cherry tomatoes, basil and lettuce.
(Courtesy of American Hydroponics, Arcata, CA).

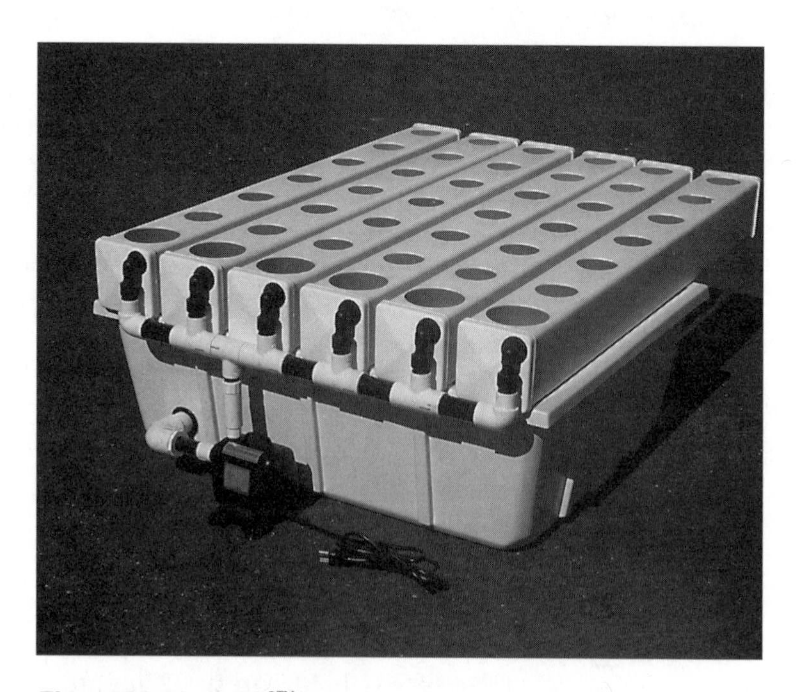

Figure 68. "AeroFlo2™" ebb-and-flow system by General Hydroponics.
(Courtesy of General Hydroponics, Sebastopol, CA).

Figure 69. This small "AeroFlo^{2TM}" unit grows 42 low profile plants such as lettuce (lower left), arugula (center), spinach, etc.
(Courtesy of General Hydroponics, Sebastopol, CA).

ent design in a combination of NFT and rock culture. They have three different models (20, 30 and 60) of what they call "AeroFlo^{2TM}" systems. These systems are recommended for small crops such as lettuce, basil, other herbs and small patio varieties of peppers and tomatoes. The different models correspond to the number of plant sites each unit contains. The AeroFlo^{2TM} 20 site system, as shown in Figure 70, includes an 8-gallon reservoir, two 6-ft. grow chambers, a pump, piping, support structure, 3" mesh pots and coco coir or "GroRox" clay pellets (Fig. 71). The grow chambers are approximately 6" by 6" so they hold up to 4 gallons of nutrient solution each making the total capacity of solution 16 gallons with the drain overflow tubes set at their high position. If the tubes are set all the way down to the base of the grow chamber, then all the solution will be retained by the reservoir only to give a total of 8 gallons. This gives the system the capability to function as deep flow with the drain tubes set high and as an NFT system with the tubes set low to the bottom of the grow chambers. They suggest using the high position when transplanting and lowering the solution level to an NFT system once the plant roots emerge into the grow chamber. The suggested retail price of this unit is $350.

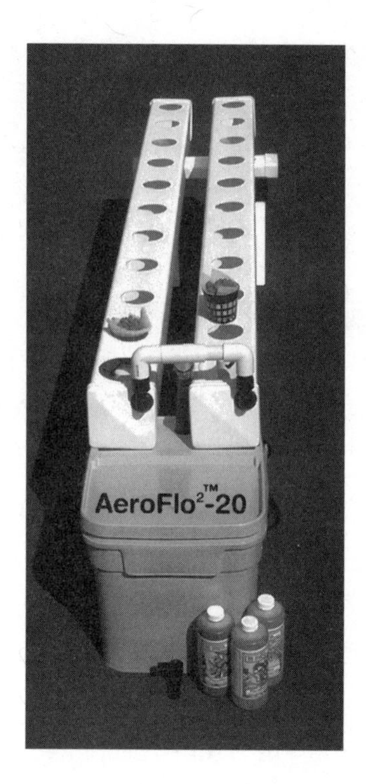

Figure 70. The "AeroFlo2™-20" has two 6-ft. growing chambers that hold 20 plants in 3-inch diameter net pots containing clay pellets or coco coir. *(Courtesy of General Hydroponics, Sebastopol, CA).*

Figure 71. "GroRox" clay pellets. *(Courtesy of General Hydroponics, Sebastopol, CA).*

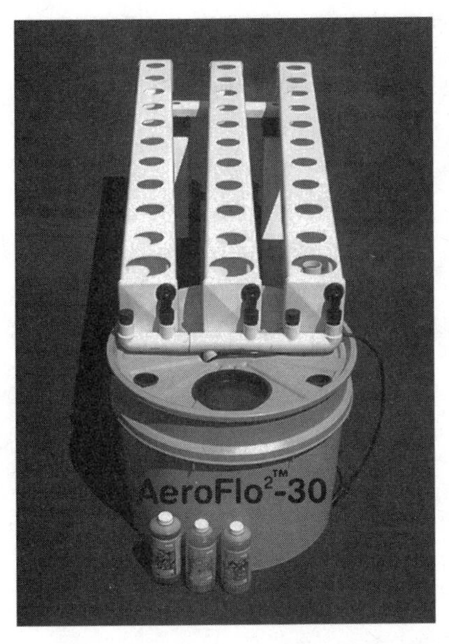

Figure 72. The "AeroFlo$^{2™}$-30" holds 30 plants in its three 6-ft. growing trays channels.
(Courtesy of General Hydroponics, Sebastopol, CA).

The AeroFlo$^{2™}$ 30 site system differs from the smaller "20" unit in that it has three 6-ft. grow chambers with a 20-gallon reservoir and uses thirty 3" mesh pots for the plants (Fig. 72). The total reservoir capacity in the flooded stage (overflow tubes set high) is approximately 34 gallons. The model "60 site system" having 6 grow chambers may be configured in two ways; the entire 6 chambers on one side for small crops such as lettuce and herbs (Figs. 73 and 74) or 3 chambers on each side of the 24-gallon reservoir for larger crops such as peppers and tomatoes (Fig. 75).

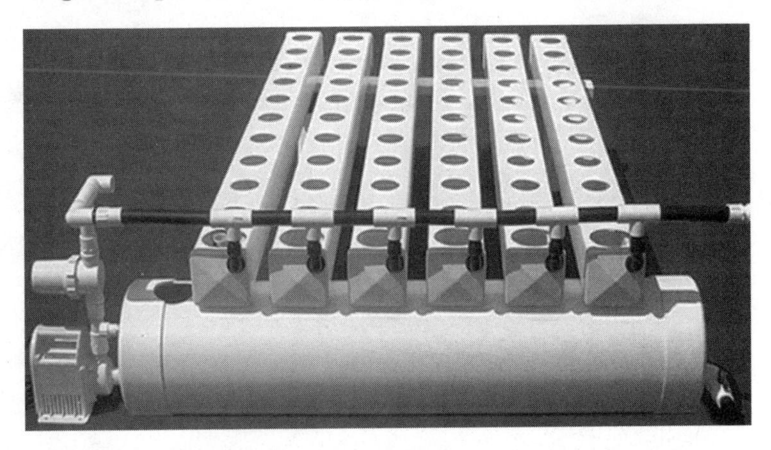

Figure 73. The "AeroFlo$^{2™}$-60" may be arranged with the six growing trays on one side of the reservoir for smaller plants as herbs and lettuce.
(Courtesy of General Hydroponics, Sebastopol, CA).

Figure 74. The nice crop of lettuce growing in the "AeroFlo^{2TM}-60" system.
(Courtesy of General Hydroponics, Sebastopol, CA).

Figure 75. Alternatively the "AeroFlo^{2TM}-60" may have 3 chambers on each side of the reservoir for taller plants such as tomatoes.
(Courtesy of General Hydroponics, Sebastopol, CA).

The total system capacity in the flooded stage is 45 gallons. Suggested retail prices for these units are $550 and $950 respectively. Two of the "60 site system" units may be connected to form 120 plant sites. This extension module costs $750.

American Agritech has an ebb-and-flow unit they call "Econojet®". It consists of two 44" x 6" x 4" flood trays that are attached to a covered 20-gallon reservoir. The trays will hold either rockwool slabs or twelve 6-inch square pots using expanded clay pebbles. Ridges in the flood trays raise the growing pots to keep the growing substrate of the pots out of the nutrient solution providing good aeration to the plant roots.

They have a series of "Micro Gardens" using ebb-and-flow systems. Basically, a growing tray sits on top of a nutrient reservoir (Fig. 76). The solution is pumped into the growing tray with pots of expanded clay pebbles (Figs. 77 & 78). Large square pots or smaller net pots may be used depending upon the plants to be grown (Fig. 79).

They also make an "Aerojet®" system, which is a combination of aeroponic and soilless cultures (Fig. 80). The grow trays are 8" wide by either 42 or 72 inches long. The 42-inch tray will hold either six 4-inch or four 6-inch

Figure 76. "Micro Garden" ebb-and-flow system basic tray on top of a reservoir by American Agritech. *(Courtesy of American Agritech, Tempe, AZ).*

Figure 77. Ebb-and-flow fill and drain pipes at the base of the growing tray of the "Micro Garden". *(Courtesy of American Agritech, Tempe, AZ).*

Figure 78. Pots are filled with expanded clay pebbles in an ebb-and-flow system. *(Courtesy of American Agritech, Tempe, AZ).*

Figure 79. Smaller net pots may be used instead of the larger square pots to hold the clay pebble substrate.
(Courtesy of American Agritech, Tempe, AZ).

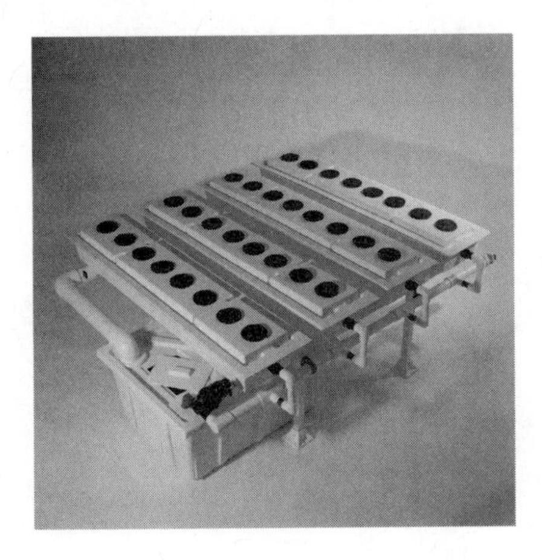

Figure 80. A four-tray "Aerojet" aeroponic system.
(Courtesy of American Agritech, Tempe, AZ).

mesh pots. Mist is applied from below in bursts of one minute on and four minutes off (Figs. 81 & 82). Fill the pots with expanded clay pebbles. A 20-gallon reservoir comes with the smaller models. Prices range from $500 to $900 for models with 2 trays to 6 trays.

American Agritech also has a series of smaller aeroponic "Micro Gardens" that use an aeroponic growing tray containing the misters (Figs. 83–85). The tray sits on top of a nutrient reservoir and has a cover to support the plants (Figs. 86, 87).

Many of these manufacturers also make propagation units. Most are an aeroponic system. These units are principally for rooting vegetative cuttings. They are particularly suitable to ornamentals. Since we start our vegetable crops from seeds as was described in Chapter 2, there is no reason to discuss these units. I only want to say that basically most are aeroponic systems such as the "RainForest" units by General Hydroponics and the "Clone Machines" of American Agritech. Depending upon their size these units are priced from $100 to $350-$400.

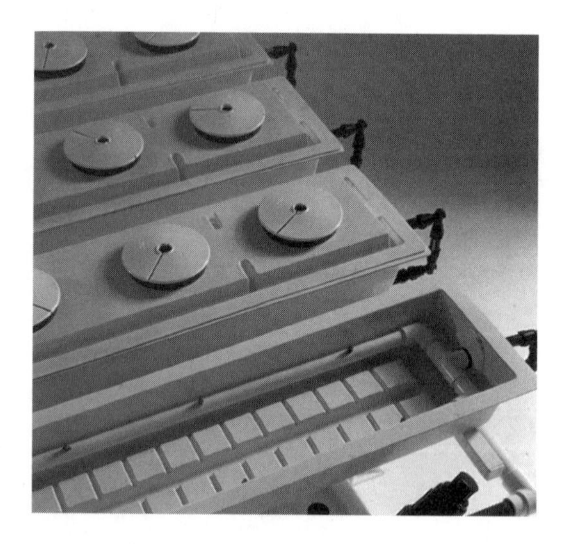

Figure 81. An "Aerojet" aeroponic system showing the mist "jets" in the base of the tray of the first tray. Note the nutrient reservoir below and the filter in the inlet line near the first tray. *(Courtesy of American Agritech, Tempe, AZ).*

Figure 82. Four-tray aeroponic "Aerojet" system with four 6-inch diameter pots per tray. *(Courtesy of American Agritech, Tempe, AZ).*

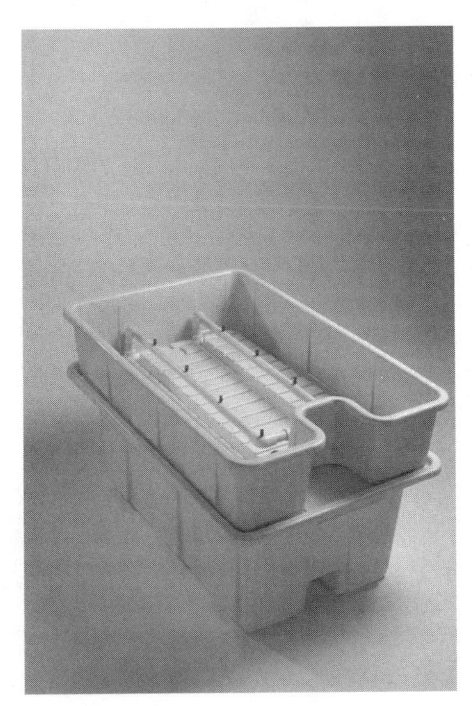

Figure 83. An aeroponic "Micro Garden" showing the mister jets at the base of the growing tray. *(Courtesy of American Agritech, Tempe, AZ).*

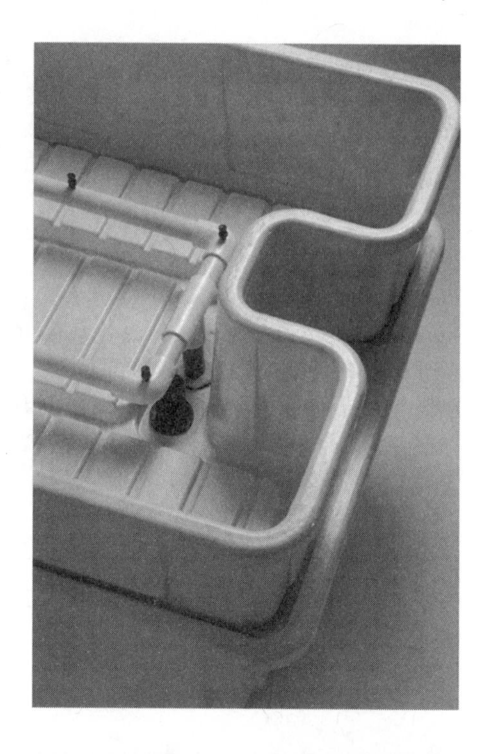

Figure 84. Close-up view of mist nozzles and plumbing from pump below in the nutrient reservoir of the "Micro Garden". *(Courtesy of American Agritech, Tempe, AZ).*

Figure 85. The aeroponic "Micro Garden" with the pots and plant support tops in place. *(Courtesy of American Agritech, Tempe, AZ).*

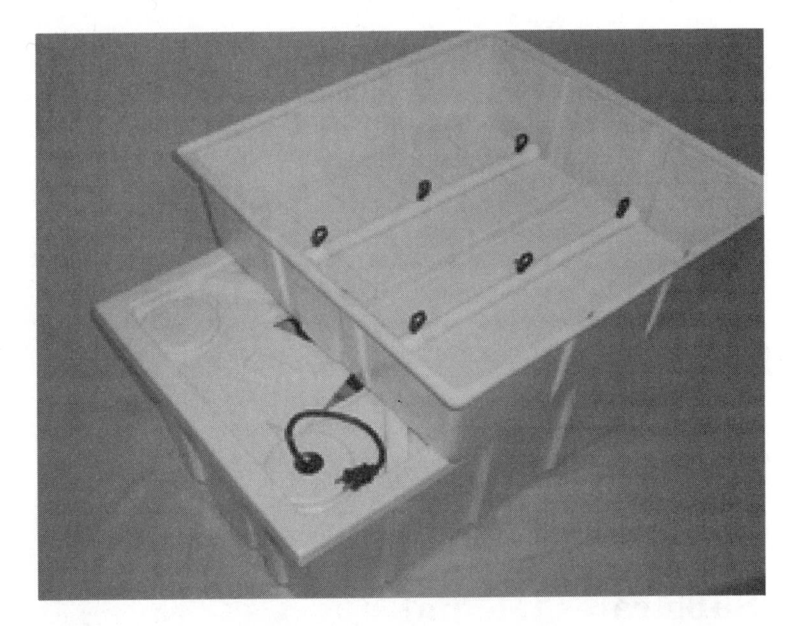

Figure 86. The “Aero 9” internal view of mist system in the base of the growing tray sitting on top of the nutrient reservoir. *(Courtesy of American Agritech, Tempe, AZ).*

Figure 87. The same “Aero 9” unit with pots and tops closed. *(Courtesy of American Agritech, Tempe, AZ).*

6 Soilless Culture

Substrates (Media)

The most popular substrates for small hydroponic units are expanded clay pebbles, rockwool and perlite. All substrates must be inert, provide adequate oxygenation and yet have good water retention as described earlier (Chapter 2). On a small-scale we do not sterilize the medium and re-use it as that is somewhat difficult and the amount of medium required per cropping period is generally fairly small and inexpensive. Many of the small commercially available hydroponic units that use mesh (net) pots are most suited to expanded clay pebbles, other small rocks or perlite. Rockwool slabs to grow mature plants will have to fit into narrow, long trays as each slab is 6" to 8" wide by 3" to 4" thick by 36" long.

Soilless Systems—Expanded Clay

Most of the combination water and soilless culture systems use expanded clay pebbles as has already been described. Various systems of soilless cultures have been designed by a number of manufacturers. Starting with the smaller one-pot systems of General Hydroponics, they have a "WaterFarm™" (Figs. 88–90), a "PowerGrower™" (Fig. 91) and an "AquaFarm™" (Fig. 92). They differ mainly in size of the growing pots and reservoirs. The WaterFarm uses a 2-gal. grow pot and a 4-gal. reservoir, whereas the AquaFarm has a 3.5-gal. grow pot and 5-gal. nutrient tank. The growing pot filled with expanded clay pebbles (Grorox) sits on top of the

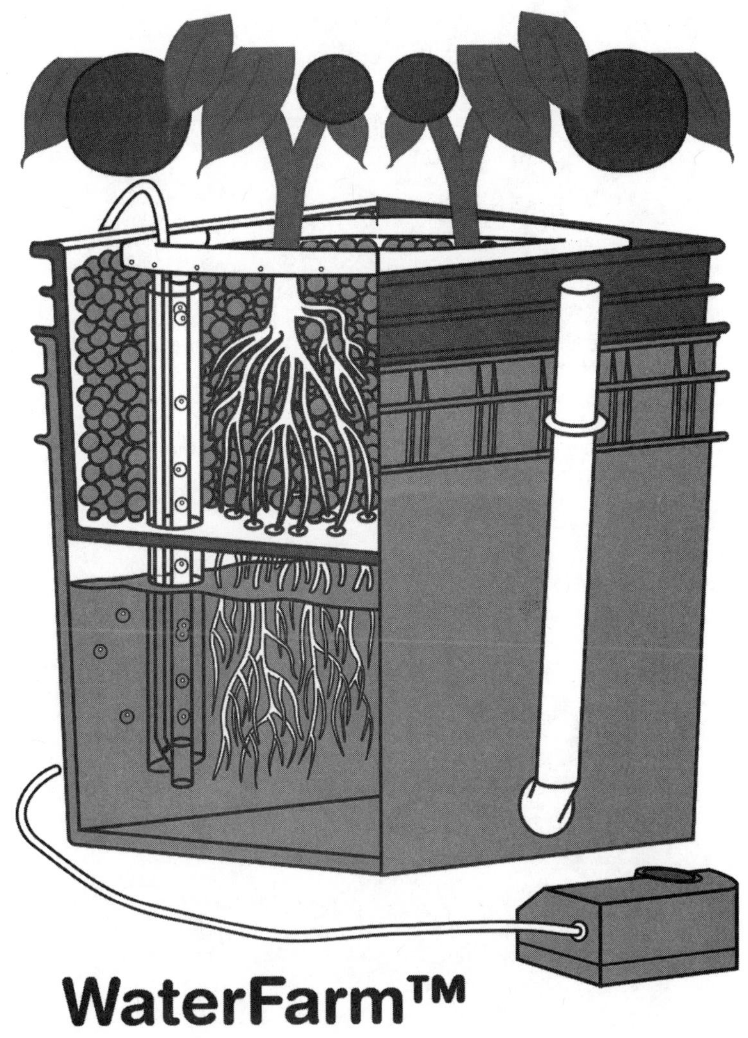

Figure 88. Diagram of "WaterFarm" one-pot system of General Hydroponics.
(Courtesy of General Hydroponics, Sebastopol, CA).

Figure 89. "WaterFarm" components include pot, reservoir, pump, tubing and clay pellet substrate.
(Courtesy of General Hydroponics, Sebastopol, CA).

Figure 90. "WaterFarm" may grow houseplants or vegetables.
(Courtesy of General Hydroponics, Sebastopol, CA).

Figure 91. Components of "PowerGrower" system.
(Courtesy of General Hydroponics, Sebastopol, CA).

Figure 92. Components of "AquaFarm" system.
(Courtesy of General Hydroponics, Sebastopol, CA).

reservoir container. The nutrient solution from the reservoir moves up by an air pump positioned outside and attached to a tube that enters a small piece of PVC pipe and is connected to a circular perforated drip ring. This system is best suited to herbs or a single plant like a dwarf patio tomato plant. It is priced at about $50.

The "PowerGrower™" may be connected in a series of pots (Fig. 93). Light Manufacturing makes a similar system of a series of pots on reservoirs they call "The Living System" (Figs. 94, 95). A basic 6-cell living system requires a space of 5 feet by 5 feet and costs about

Figure 93. The "PowerGrower" may be connected in series to grow a number of plants. Nutrient reservoir is on the right.
(Courtesy of General Hydroponics, Sebastopol, CA).

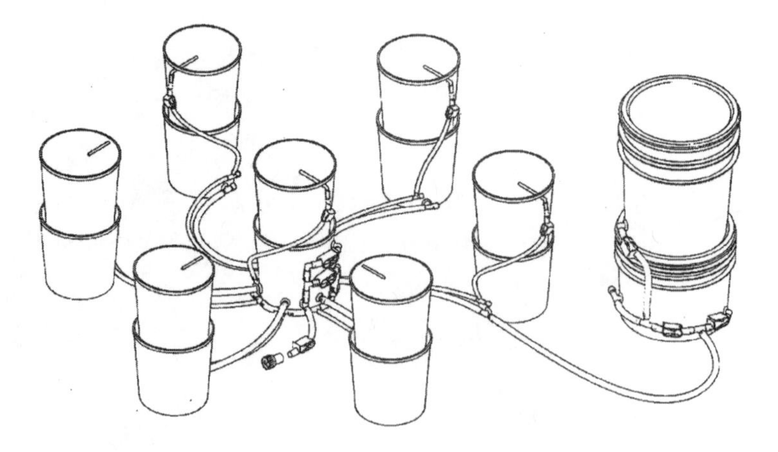

Figure 94. Diagram of the layout of "The Living System" by Light Manufacturing Co.
(Courtesy of Light Manufacturing Co., Portland, OR).

Figure 95. A six-cell pot-on-pot unit of "Living System" requires a space of 5 ft by 5 ft.
(Courtesy of Light Manufacturing Co., Portland, OR).

$180. These "Living Systems" may be expanded in multiples of 6 cells. Tomatoes and peppers may be grown as one per pot allowing 4 square feet per plant.

General Hydroponics have a somewhat larger unit they call "Eve's Garden®". This system consists of a series of 6 or 12 plastic "Dutch Bato Bucket" pots (Fig. 96) that sit on a return catchment gutter (Fig. 97). Each bato bucket has a siphon drain fitting that maintains a water level of about 1-inch at the bottom of the pot (Fig. 97). It drains the pot to the underlying return on which the bucket is partially sitting. The catchment gutter returns the solution to the nutrient tank below the supporting frame on which the buckets sit (Fig. 98). A submersible pump feeds the plants with a drip irrigation line to each pot. The pots are filled with a mixture of Grorox and coco coir. Vegetables and flowers grow well in this system (Fig. 99).

I have good success with growing tomatoes, peppers,

Figure 96. Cross-section diagram of "Dutch bato bucket" pot containing clay pellets.
(Courtesy of General Hydroponics, Sebastopol, CA).

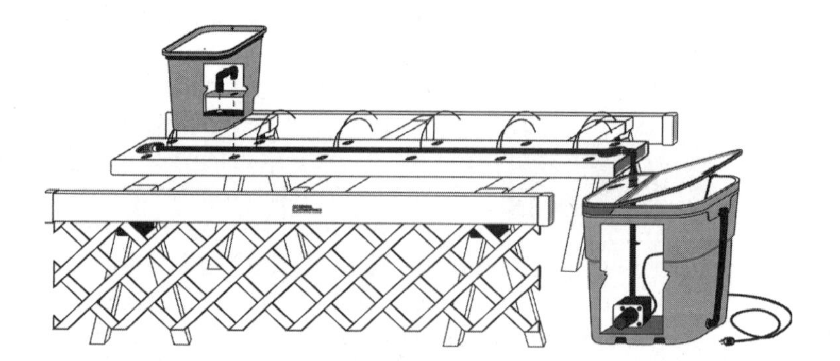

Figure 97. Diagram of "Eve's Garden" set up with Dutch pots on a supporting frame with drain channels, drip irrigation system and nutrient reservoir with pump.
(Courtesy of General Hydroponics, Sebastopol, CA).

Figure 98. Six-pot system of "Eve's Garden".
(Courtesy of General Hydroponics, Sebastopol, CA).

Figure 99. Vegetables growing in 12-pot system of "Eve's Garden". Zucchini, tomatoes, eggplant and melons.
(Courtesy of General Hydroponics, Sebastopol, CA).

Figure 100. American Agritech "Econo-Jet" ebb-and-flow hydrogarden tray sits on top of a 20-gallon nutrient tank. *(Courtesy of American Agritech, Tempe, AZ).*

and cucumbers commercially in bato buckets with a perlite substrate. They all grow very well with perlite, so I think that it would be better to use perlite instead of the expanded clay pebbles mixed with coco coir. Just be careful with the spacing for these plants as we discussed a number of times earlier. The suggested prices for these units are $450 and $550 respectively for the 6- and 12-pot systems.

American Agritech has an "Econo-Jet" series of ebb-and-flow hydrogardens that may use expanded clay pebbles, rockwool, perlite or a coco coir mixture as the substrate. The grow tray measures 44" x 24" x 7" deep. It sits on top of a 20-gal. nutrient reservoir as shown in Figure 100. It is priced at just under $200. A "Jetflo®" model is similar to the "Econo-Jet" one, except that it is modular in that several trays may be supported above the nutrient reservoir (Fig. 101). It sells for $530 and a 3-tray system for $720. With these models you may fill

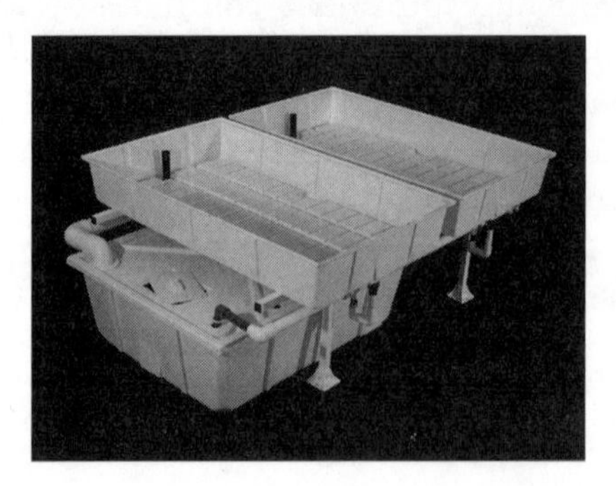

Figure 101. A "Jetflo" model is similar to the "Econo-Jet" unit except that it is modular with several growing trays may be set on a supporting frame above the nutrient reservoir. *(Courtesy of American Agritech, Tempe, AZ).*

the trays with a substrate or I prefer to use 5-inch or 6-inch diameter mesh pots to be certain of getting adequate drainage. Again the number of plants is not really determined by the area of the trays, but by the floor area when growing vine crops.

Rockwool Culture

With rockwool culture start the seedlings in rockwool cubes and transplant to rockwool blocks before transplanting a second time to the rockwool slabs. Setting up your own rockwool system is fairly simple. You need a 20- to 30-gallon solution tank, support structure, pump and drip irrigation lines, PVC piping, and a series of trays in which to place the rockwool slabs. The trays should be at least 8 inches wide by a multiple of 3 feet as the rockwool slabs are approximately 3 ft long.

American Agritech sells what they term "plumbed grow trays" for rockwool or ebb-and-flow systems. They have 44" x 6" x 4" and 42" x 8" x 4" trays complete with plumbing fittings for drainage and drip irrigation. These hold one slab or six 6-inch square pots. They also have 44" x 6" x 4" and 42" x 8" x 4" ebb-and-flow trays. Purchase these trays for your system at from $20 to $30 each and build the rest. You can buy a small storage box (available in K-Mart or Wal-Mart) to serve as the nutrient reservoir. Pumps and irrigation supplies are available in many garden centers, hydroponic shops and irrigation stores.

Alternatively, you may purchase complete systems such as those available from American Agritech. They have a number of models under their "Jetstream®" category. Four-tray and six-tray systems are available in 6", 8" or 9" widths by 42- to 44-inches long (Figs. 102, 103). Each tray contains one rockwool slab. A submersible pump feeds the individual plants in their rockwool blocks with a drip irrigation line via a PVC header. A 2-inch diameter PVC catchment pipe returns the solution to the 20-gal. reservoir. Four plants are set

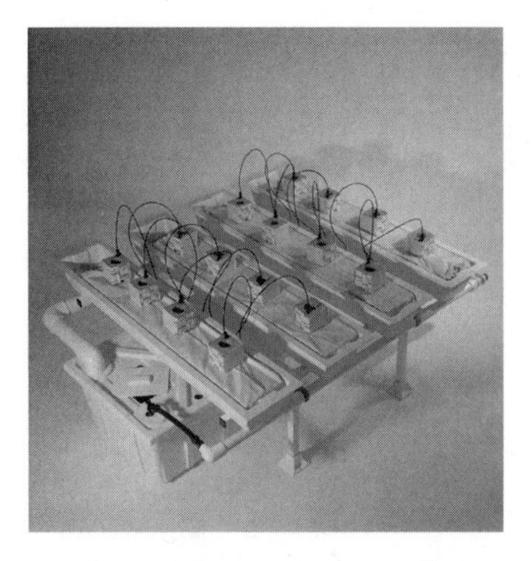

Figure 102. American Agritech "Jetstream" rockwool culture unit four-tray system. Note the use of the drip lines to each rockwool block. *(Courtesy of American Agritech, Tempe, AZ).*

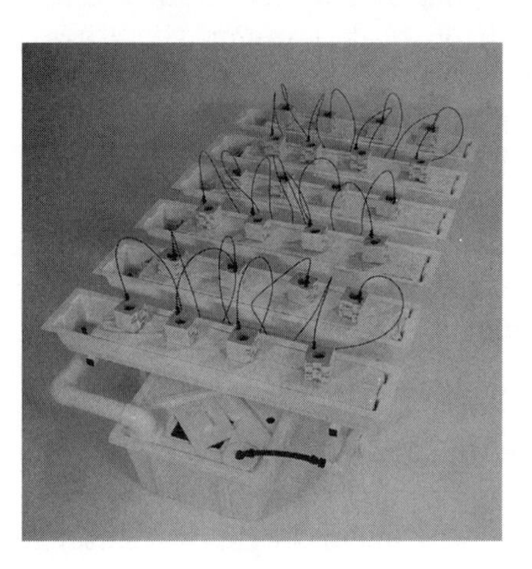

Figure 103. Six-tray top feed "Jetstream" rockwool culture unit. *(Courtesy of American Agritech, Tempe, AZ).*

onto each rockwool slab in each grow tray. A four-tray system has 16 plant sites and a 6-tray one has 24 sites, but always remember the spacing rule. This rockwool system is ideal for most vine crops, just be careful with the area per plant needed. Prices range from $350 for a 2-tray unit to $790 for an 8-tray one. These units are easily converted to mesh pots with expanded clay pebbles or perlite using the same system of drip irrigation.

For the hydrogardener who is just starting out, American Agritech has designed a smaller version they call the "Jetstream Mini®". The grow trays are 24 inches long by 8 inches wide. There are four different sizes, 2-tray, 4-tray, 6-tray and 8-tray units. The grow

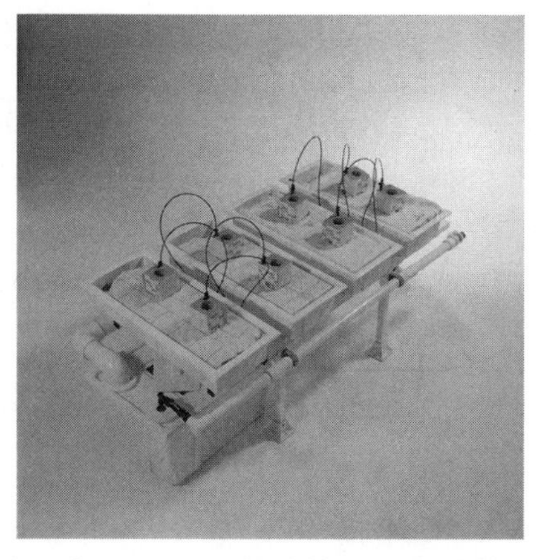

Figure 104. "Jetstream Mini" has 4 trays supported above a 10-gallon nutrient reservoir. It holds 8 plants. *(Courtesy of American Agritech, Tempe, AZ).*

trays are supported above a 10-gal. nutrient reservoir as shown in Figure 104 of a 4-tray system. Each tray will hold two plants. The rockwool culture unit holds one-half of a standard slab in each tray. Each plant is fed by a drip line and emitter from a submersible pump attached to a PVC header. The returning solution drains back to the reservoir via a 2-inch diameter PVC pipe. The trays are also available with an optional cover that holds two, 5-inch mesh pots with expanded clay pebbles or perlite substrate. This unit is easily expanded in the number of grow trays. There is also a "Jetflo Mini®" using a 22" x 22" x 7" ebb-and-flow tray mounted to a 20-gallon reservoir (Fig. 105). It will hold six plants. I feel that this system is well designed to grow vine crops. Just be careful with your spacing of these crops. Remember to V-cordon train them away from the

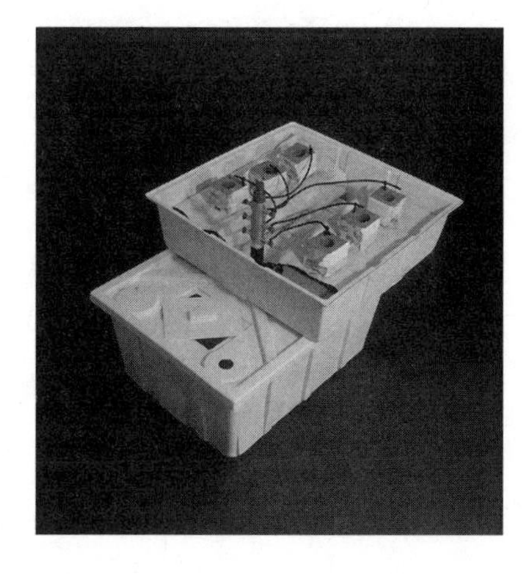

Figure 105. A "Jetflo Mini" uses a ebb-and-flow tray mounted to a 20-gallon reservoir. It holds 6 plants fed by a drip irrigation, rockwool culture system. These units are interchangeable for growing as ebb-and-flow or rockwool by some simple modifications using available conversion kits they term "AquaShuttle". *(Courtesy of American Agritech, Tempe, AZ).*

grow trays to obtain sufficient light for the plants.

The manufacturer's suggested retail prices are $330, $430, $520 and $600 for the various tray models starting from the 2-tray through the 8-tray models.

Perlite Culture–Vertical Plant Towers

I have been very successful in growing lettuce, bok choy, herbs and strawberries with plant towers. The substrate of choice is perlite. You may wish to mix 15% coco coir with it for higher water retention. I have not found that necessary, in fact mixing peat or coco coir with the perlite will slow drainage from one pot to the next and sometimes causes overflowing of the nutrient solution.

Verti-Gro, Inc. of Lady Lake, Florida developed over the past 5 years a unique system of vertical culture using Styrofoam pots stacked one on top of the other to a height of 10 pots. The culture was developed for the growing of strawberries in greenhouses. With these towers the plant density could be increased up to 8 times what would be possible when growing them in normal horizontal beds. They soon became aware that similar low-profile crops such as spinach, herbs and lettuce adapted readily to this system of growing.

The Styrofoam pots are stacked one above the other by rotating them by 45 degrees so that the four corners of each pot is exposed as shown in Figure 106. The Styrofoam pots measure 9" x 9" x 8" deep with slightly tapered sides. Their top lip is notched on each side to fit the bottom of the adjoining pot (Fig. 106). Here is how they are set up. A collection pan sits on a drainpipe (Fig. 107) to collect the drain water and return it to a solution reservoir. An indoor unit for your house would need to be raised on a support frame to keep the drainpipe at an elevation above the level of the nutrient reservoir. Fill the collection pan with rocks such as the expanded clay pebbles. While filling the pan, place a small 3/4" diameter tee with 4-inch pieces attached to each opening in the bottom of the pan. This will stabi-

Figure 106. Plant tower formed by stacking Styrofoam pots one on top of the other by turning them at 45 degrees to each other. Notches fit each pot in its exact position.

lize the support bar that keeps the pots from falling over. A piece of galvanized steel electrical conduit is placed into the upper end of the tee. A 3/4"–diameter PVC pipe slides over the conduit to act as a sleeve over which the pots are threaded through a central hole in each pot (Fig. 108). A swivel plate between the top of the tee extension and the first pot allows the entire tower to be rotated every day to better distribute the light entering the crop (Fig. 107).

In your house you will probably only be able to stack 8 pots high, as space is required for the lights above. Slide all the pots onto the vertical pipe support before filling them with the perlite substrate. Then, level the support pipe in all directions before attaching the top of it to a support wire strung across the room. Once the pots are supported on the central pipe, fill each pot starting from the bottom one with perlite up to within 1 1/2" of the rim. Allowing a bit of air space between each pot will help to prevent roots growing

Figure 107. Collection pot sits on drainpipe to retrieve all drainage. Swivel plate allows the plant tower to be turned easily each day to get uniform light on all of the plants.

Figure 108. Electrical conduit with plastic pipe sleeve supports pots vertically from an overhead support wire. Various herbs growing in plant towers of 10 pots.
(Courtesy of Cuisinart Resort & Spa, Anguilla, B.W.I.).

from one pot down into the next through the perforated bottoms of the pots. The next step is to install a drip irrigation system. Using a submersible pump in a nutrient reservoir, attach a PVC pipe vertically to the upper height of the tower and then convert it to 1/2-inch black polyethylene hose running along the support wire to all towers. Insert at least three drip lines to each tower, two in the top pot and another one halfway down the tower. This will assist in distributing the solution evenly down the plant tower (Fig. 109). Each plant tower requires 10 square feet of floor area to permit adequate light penetration to the crop.

These towers are very productive (Figs. 110, 111).

Figure 109. Two drip lines enter the top pot and a third one is located in the middle of the plant tower with thyme. *(Courtesy of Cuisinart Resort & Spa, Anguilla, B.W.I.).*

One tower of 8 pots will produce 32 head of lettuce or bok choy monthly if you start the seedlings in rockwool cubes and transplant them into the plant tower at 21 days (Figs. 112–114). Sequence your planting so that you may harvest every day or so depending upon your personal needs. Herbs can be sown directly into the pots of the plant towers. Herbs will take several months to become well established (Fig. 115). After that you may cut them daily, as a little trimming each day is better for their continued production than to cut them back severely once a week or so. To grow strawberries you need to purchase live plants as was described earlier in Chapter 2. Place one plant in each corner of the plant tower pots. They will take several months to start producing. Growing vine crops in plant towers is not very convenient as only one plant could be grown in a single pot near the bottom of the tower. We have already talked about many more suitable systems for growing your tomatoes, peppers and cucumbers.

Figure 110. Italian parsley growing in plant tower. *(Courtesy of Cuisinart Resort & Spa, Anguilla, B.W.I.).*

Figure 111. Chives in plant tower.
Courtesy of Cuisinart Resort & Spa, Anguilla, B.W.I.).

Figure 112. Thyme in plant tower. Note the plastic pipe sleeve on conduit support pipe at the top of the plant tower.
(Courtesy of Cuisinart Resort & Spa, Anguilla, B.W.I.).

Figure 113. Bok Choy in plant tower. *(Courtesy of Cuisinart Resort & Spa, Anguilla, B.W.I.).*

Figure 114. Romaine lettuce in plant tower. *(Courtesy of Cuisinart Resort & Spa, Anguilla, B.W.I.).*

Figure 115. Mint, oregano and Italian parsley (right to left) in plant towers. *Courtesy of Cuisinart Resort & Spa, Anguilla, B.W.I.).*

Hobby Plant Towers

Verti-Gro, Inc. also has a number of small self-contained hobby units should you prefer to purchase a complete system that can be easily assembled within an hour. The two most popular models are two-stack and three-stack systems. The pots in these systems stack on top of a nutrient reservoir. The two-stack is 5 pots and the three-stack 7 pots high (Figs. 116, 117). They also have a 4-stack system that is 8 pots high for a total of 32 pots.

The 2-stack "Vertical Garden" comes with 10 pots, submersible pump, timer, pipes and fittings, 12.5-gal. tank, perlite and nutrients. The nutrient solution is pumped up a 3/4" diameter PVC pipe passing through the center of each pot. An adjustable emitter at the top regulates the flow rate (Fig. 118). The solution is pumped every few hours for 15 minutes operated by a small time clock. The solution percolates from the top pot down through the lower pots and returns to the

Figure 116. Two-stack, hobby plant tower. Basil planted in top pots, viola on left and marigolds on right.
(Courtesy of Cuisinart Resort & Spa, Anguilla, B.W.I.).

Figure 117. Lettuce growing in three-stack, hobby plant tower. Note reservoir at bottom on which the plant towers sit. Pipe in center is to support the towers. *(Courtesy of Cuisinart Resort & Spa, Anguilla, B.W.I.).*

Figure 118. Irrigation bubbler at top of each plant tower stack. A timer-operated pump initiates irrigation cycles.
(Courtesy of Cuisinart Resort & Spa, Anguilla, B.W.I.).

reservoir below on which the pots sit. This unit will grow 40 lettuces or a combination of different herbs and lettuce.This unit is priced at about $200.

The 3-stack system comes with 21 pots and the rest of the components as with the 2-stack unit. A triangular base reservoir holds a larger volume of solution (about 25 gallons) than the smaller rectangular base of the 2-stack garden. It will grow 84 head of lettuce or a combination of herbs and other crops such as spinach, bok choy, lettuce, arugula, chard or flowers (Figs. 117, 119). These plant towers have the flexibility of growing different crops in each pot. For example, I have grown 4 basil plants in the top pot as they require more space; oregano, thyme, parsley, chives, sweet marjoram, chervil, mint, cilantro and others in the remaining pots. Grow a selection of these herbs to meet your personal consumption. This unit costs about $400. A 4-stack unit costs about $450.

Figure 119. Bibb lettuce in left tower, beet tops in middle and oakleaf lettuce in right tower. *(Courtesy of Cuisinart Resort & Spa, Anguilla, B.W.I.).*

Popular Hydroponics

This is a term coined by a group of non-profit organizations from the universities in Colombia and Peru to get people involved in producing their own vegetables in poor communities. The universities together with community members developed very inexpensive, simple, non-mechanical methods of growing hydroponically on a small scale. The principle is to use inexpensive local materials or waste such as old tires or plastic containers. Drainage holes are drilled in the containers prior to filling them with media such as peat, rocks, rice hulls, coco coir or mixtures of these substrates. Seeds are sown in these gardens and the plants fed with a hydroponic nutrient solution developed by the universities. They are given seeds, nutrients and other supplies at a very low cost. Lettuce, beets, chard, strawberries, herbs, basil, peppers, tomatoes and local crops of their normal diet are grown (Fig. 120). In addition, they grow grasses as fodder for their animals (Fig. 121).

They also use simple systems of column culture with plastic sacks containing a mixture of rice hulls and coco coir. A container with the nutrient solution is attached to the top of the sack where it percolates slowly into the sack. The drainage runs into a collection pan at the bottom so that the solution may be poured back into the container at the top for use again (Fig. 122). Strawberries are popular in this system. You can construct a simple raft culture system from lumber, polyethylene and Styrofoam. Many poor communities in Latin America construct such units to grow lettuce, basil, herbs, chard, spinach and other leafy crops (Fig. 123). Cut a 4-ft x 8-ft Styrofoam sheet in half to get

Figure 120. Inexpensive sack culture of growing lettuce, herbs and strawberries in poor communities of Peru.

Figure 121. Grasses growing for cattle in inexpensive containerized hydroponics containing mixtures of rock, sand, peat, rice hulls and coco coir. This is "Popular hydroponics" in Peru.

Figure 122. Sack culture in Peru using a mixture of coco coir and rice hulls. Note collection pan at bottom and nutrient reservoir at top of each sack.

Figure 123. Inexpensive raft culture system for growing lettuce and herbs in poor communities of Peru.

exactly 4 ft by 4 ft. Cut sixty-four 3/4"–diameter holes at 6" by 6" centers with a hole saw for the plant sites. The Styrofoam panel supports the plants and insulates the solution below in the reservoir from light and heat. Make a wooden or brick frame at least one-inch wider in the inside dimensions (49" x 49"). This gives you room for the Styrofoam to be easily taken in and out of the reservoir. The height should be about 7 to 8 inches. Line the reservoir with double 10-mil polyethylene or a 20-mil swimming pool vinyl. If you have time to agitate the nutrient solution with a whisk 3 to 4 times a day, you can get by without an aeration pump. The easiest way to aerate the solution is to use an aquarium pump attached to air stones in the reservoir. This do-it-yourself project should cost under $50.

While I present these systems here as all manually

operated, you can use the same principles to construct such systems and automate them with a tank, submersible pump and irrigation components. The other point I wish to make here is that hydroponics can be practical for many different system designs from sophisticated, automated to simple, manual ones. So, it is applicable to all income levels and backgrounds. You do not need to be a rocket scientist to successfully grow hydroponically, just be able to follow instructions to best achieve the results you want for your particular level of growing. These simple systems used for poor communities may be the answer for many of these people currently under malnutrition to better provide basic nutrition for their families. They often live in arid regions where the soils are very poor in nutrition and structure compounded with the scarcity of water. In this situation hydroponics is the answer to growing fresh vegetables.

7 Hobby Hydroponic Supplies and Information

Hydroponic Supplies

There are over 745 hydroponic stores worldwide listed on the Internet. I am sure there are others in addition to these. Also, related industries such as hardware and builders' supplies, irrigation suppliers, garden centers, fertilizer companies and discount stores sell components needed, especially for the do-it-yourself hobbyist. The Web site *www.hydromall.com* lists 4 stores in Asia, 169 in Australia, 171 in Canada, 67 in Europe, 7 in New Zealand and 314 in the USA with 13 online stores. The four stores in Asia are in Singapore. There are no listings of those in Japan, Hong Kong or Malaysia where there must be many more as hydroponics is well known in those countries. The distribution of the shops in Canada is: Alberta (8), British Columbia (78), Manitoba (7), New Brunswick (2), Nova Scotia (1), Ontario (41), Quebec (30), Saskatchewan (3) and Yukon Territory (1). Here are the locations in Europe by country: Austria (2), France (2), Germany (4), Holland (1), Ireland (1), Italy (3), Spain (8) and United Kingdom (46). I expect that Holland has an abundance of stores as Holland has the largest commercial hydroponic greenhouse industry in the world. There are stores listed for 39 states in the USA with the majority in California (70).

Most of these stores have their own Web sites that you can browse to explore what products they offer. In addition, there are the Web sites of some of the manufacturers of hydroponic equipment. The following Web sites include the manufacturers of units described earlier in this book.

Table 4. Web sites of Manufacturers of Hydroponic Units and Accessories.

Company	Web site
American Agritech	*www.americanagritech.com*
American Hydroponics	*www.amhydro.com*
North American Greenhouse Supplies	*www.greenhousesolutions.com*
Diamond Lights	*www.diamondlights.com*
General Hydroponics	*www.generalhydroponics.com*
Homegrown Hydroponics	*www.hydroponics.com*
Light Manufacturing Company	*www.litemanu.com*
Stocker Hort. & Hydroponics Supplies	*www.hydroponics.co.nz*
North American Hydroponics	*www.nahydroponics.com*
P.L. Light Systems	*www.pllight.com*
Verti-Gro, Inc.	*www.vertigro.com*
CropKing, Inc.	*www.cropking.com*

Many Web sites like “HydroMall” give links to other websites. HydroMall has over 720 links to information related to hydroponics. There are over 1750 fruit and vegetable links. This website has diverse information with articles on growing, products, services, stores, e-mail addresses, advertising and much more.

Hydroponic Services

A lot of information is available through private and governmental Web sites. Such Web sites offer information on pest and disease identification and control, cultural techniques for crops, nutrition, nutritional disorders and remedies, hydroponic systems, seeds and new cultivars most suitable to hydroponic growing. Some of these sites are:

Table 5. **Web sites of Government & Universities Offering Information on Hydroponics.**

Government and Universities-Hydroponics
www.aceis.agr.ca
www.ag.arizona.edu/hydroponictomatoes
www.cals.cornell.edu/dept/flori/cea/programs.html
www.ext.msstate.edu:80/anr/plantsoil/vegfruit/tomato/ghtomato/faq.html
www.greenhouseinfo.com
www.growroom.com
www.hort.purdue.edu/newcrop/home
www.usda.gov
www.ontariogreenhouse.com

Table 5. Web sites of Government and Universities Offering Information on Hydroponics. ***(cont'd)***

Government and Universities-Hydroponics

www.agf.gov.bc.ca/croplive/cropprot/prodguide.html

www.ces.ncsu.edu/greenhouse-veg/

http://res.agr.ca/harrow/

http://res2.agr.ca/harrow/bk/tom-toc.htm

www.orst.edu/Dept/NWREC/tomatogh.html

http://nj-nscort.rutgers.edu/visitor/tps/index.html

These are a few sites offering technical information on horticulture and hydroponics. Most have links to other sites on specific topics. Some are very extensive and give references to their publications that you may obtain and may recommend other journals and articles on related topics.

Table 6. Web sites on Identification and Control of Pests & Diseases Using Integrated Pest Management (IPM).

Pests & Diseases-IPM

www.koppert.nl

www.anbp.org

www.intertechserv.com

www.biobest.be

Table 6. **Web sites on Identification and Control of Pests & Diseases Using Integrated Pest Management (IPM).**

Pests & Diseases-IPM

www.mycotech.com

www.bioworksbiocontrol.com

www.nysaes.cornell.edu/ent/biocontrol/

www.cas.psu.edu/docs/cashome/progper/greenhouse.html

www.bugstore.com

www.insectuary.com

www.ipm.ucdavis.edu/PMG/PESTNOTES/pn002.html

These sites on integrated pest management (IPM) have many color photos of pests, diseases and beneficial insects and microorganisms that control them. The sites are useful in identification of any pest problem you may encounter. Full descriptions of both the pests and beneficials are presented that assist you in determining what control measures are available.

Table 7. **Web sites of Seed Houses.**

Seed Houses -Varieties	
De Ruiter Seeds Inc.	www.deruiterusa.com
Johnny's Selected Seeds	www.johnnyseeds.com
Ornamental Edibles	www.ornamentaledibles.com

Table 7. Web sites of Seed Houses.

Seed Houses -Varieties	
Richters Herbs	www.richters.com
Stokes Seeds Ltd.	www.stokeseeds.com
Rijk Zwaan USA	www.rijkzwaan.nl

Seed catalogs are very informative. They indicate the kind of conditions the plants like for their growth. They describe the characteristics of the vegetables, cropping periods, number of seeds per unit weight and other interesting aspects of all varieties. In addition, it is nice to see what your ideal fruits and vegetables should look like. Seed catalogs also introduce new varieties every year with extensive descriptions of their characteristics. While often these catalogs are available online, it is entertaining to spend some leisure time studying the catalogs in printed form.

Hydroponic Organizations and Internet Chat Clubs

There are a number of hydroponic societies that promote new technology and products. They generally have annual meetings or conferences at which they obtain experts within the field of hydroponics to give presentations. I have found these very informative and a pleasure to meet people like you who have been inspired by hydroponics. It is always a learning experience to see new research and products. Most conferences have a Hydroponic Suppliers' Trade Show displaying products offered by companies.

Table 8. Addresses and Web sites/e-mails for Hydroponic Societies.

Address	Web site/e-mail
Hydroponic Society of America (HSA) P.O. Box 1183 El Cerrito, CA 94530	www.hydroponics.org
Hydroponic Merchants Association (HMA) 10210 Leatherleaf Court Manassas, VA 20111-4245 USA	www.hydromerchants.org
International Society for Soilless Culture (ISOSC) Secretariat of ISOSC P.O. Box 52 Wageningen, Netherlands	
Australian Hydroponic & Greenhouse Association (AHGA) AHGA Administrator, P.O. Box 538 Narrabeen, NSW, 2101. Australia	www.ahga.org.au
Singapore Society for Soilless Culture Block 461 #13-75, Crawford Lane, Singapore	http://soillessculture.bizcal.com
Asociacion Hidroponica Mexicana A.C.	www.hidroponia.org.mx
Centro de Investigacion de Hidroponia y Nutricion Mineral Univ. Nacional Agraria La Molina Av. La Universidad s/n La Molina, Lima 12. Peru	www.lamolina.edu.pe/hidroponia
Informative Bulletin (bimonthly)	www.lamolina.edu.pe/hidroponia/red.htm

There are several hydroponic forums where you may sign up to be part of discussion groups online. You may submit questions for advice from other growers and hydroponic experts. This is also a good method of keeping informed of new products. Send an e-mail to the site to sign up as a member.Here are two:

Table 9. Web sites/e-mails of Hydroponic Forums.

hydroforum@fesersoft.com

hydrolist@hydroponics.org

http://forums.gardenweb.com/forums/hydro/

www.hydrohangout.com

www.hobbyhydro.com

Hydroponic Magazines

These popular magazines are a must for any hydroponic grower. They have both popular and technical articles. Both are published bimonthly.

Table 10. Addresses and Web sites of Hydroponic Magazines.

Address	Web site
Practical Hydroponics & Greenhouses Casper Publications Pty Ltd. P.O. Box 225 Narrabeen, 2101 Australia	*www.hydroponics.com.au*

Table 10. Addresses and Web sites of Hydroponic Magazines.

The Growing Edge Magazine P.O. Box 1027 Corvallis, OR 97339	*www.growingedge.com*

These magazines also have extensive advertising by manufacturers and suppliers of hydroponic products to keep you informed of new developments.

References

Many books are available on hydroponics. Books are sold by hydroponic stores, garden centers and on the Internet such as *www.amazon.com* and *www.barnesandnoble.com*. The Hydroponic Society of America, The Growing Edge and Practical Hydroponics and Greenhouses also sell books.

AHA Conference Proceedings

Basic Hydroponics (2nd Ed.)—Ed Muckle

Beginning Hydroponics: Soilless Gardening—
Richard E. Nicholls

Cultivos Hidroponicos (5th Ed.) —
Howard M. Resh, PhD

Gardening Indoors—George F. Van Patten, et al.

Gardening Indoors with CO2—
George F. Van Patten, et al.

Gardening Indoors with HID Lights—
George F. Van Patten, et al.

Gardening Indoors with Rockwool—
George F. Van Patten, et al.

Greenhouses for Homeowners and Gardeners—
Natural Resource, Agriculture & Engineering Science

Home Hydroponic Gardens—Peggy Bradley

Hydroponic Crop Production—Lon Dalton &
Rob Smith

Hydroponic Food Production (6th Ed.) —
Howard M. Resh, PhD

Hydroponics for the Home Gardener—
Stewart Kenyon

Hydroponic Gardening—Raymond Bridwell

Hydroponic Gardening—Lon Dalton & Rob Smith

Hydroponic Home Food Gardens—
Howard M. Resh, PhD

Hydroponic Hot House—James B. DeKorne

Hydroponic Lettuce Production—Dr. Lynette Morgan

HSA Annual Conference Proceedings

Hydroponic Tomato Production—Jack Ross

Hydroponic Tomatoes —Howard M. Resh, PhD

Hydroponically Speaking—CO2 Enrichment—
Steven Carruthers

Hydroponically Speaking—Lighting—Steven
Carruthers

Hydroponics for Everyone—Dr. Struan K. Sutherland

Hydroponics Questions & Answers—
Howard M. Resh, PhD

Knowing and Recognizing—M. Malais and
W.J. Ravensberg

Manual Practico de Hidroponia—
Alfredo Rodriguez Delfin, C. Milagros,
Marilu Hoyos and Fernando Falcon

Natural Pest Contol—Andrew Lopez

The ABC of NFT—Dr. Allen Cooper

The Best of The Growing Edge—Tom Alexander

The Handbook of Hydroponic Nutrient Solutions—
Carl Berry

Closing Comments

I wish to share with you some comments received over the years from hydroponic enthusiasts. I became involved in hydroponics while a graduate student at The University of British Columbia in Vancouver, B.C., Canada. While the initial work was in commercial hydroponic greenhouses, it soon became evident to me that there was a great potential for this type of growing at the hobby level. An engineer and I incorporated a company to start building backyard hydroponic greenhouses. We used to go to exhibitions to display our hobby greenhouses and to our surprise found a constant line-up at our exhibit to enter our greenhouse and see tomatoes, cucumbers and lettuce growing in rocks. Many people would come in disbelief that these plants were real and they would have to feel them before becoming convinced that they were not plastic. That was back in 1975 when hydroponics was known in the commercial greenhouse industry, but not as a hobby.

We began manufacturing small greenhouses of a basic size of 10½ ft. by 12 ft. (Fig. 124). Most clients thought they were fairly large and would not need all the space, but within months once they were producing their own vegetables they became very diversified in their crops, and of course there were always those houseplants that needed rejuvenating in the greenhouse. Soon their greenhouse was so full they ran out of space. I always remember the biggest complaint we would soon get was "I wish you would have sold me a larger one". The same is true of small indoor units.

Figure 124. Small 10 ½ ft by 16 ft backyard greenhouse.

Once you get fresh vegetables in the middle of the winter, you too will say, "I must expand this hydroponic garden". Our clients found the greenhouse was a very relaxing place to escape from their everyday stresses. They could enter the greenhouse in the middle of a dark rainy day in January and be in a tropical paradise with their plants growing under artificial lights. So, there are many benefits of hydroponic growing from pleasure of working with your plants, escape from stress, to a rewarding experience with garden fresh tomatoes, cucumbers, lettuce and herbs in the height of winter.

My purpose with this book is to show you the various types of hydroponic indoor garden units that are prefabricated and easily set up for operation immediately. I realize that all such units cannot be covered in a single book, so the aim is to provide you with a cross-section of typical units available under the various hydroponic cultural methods. Most units are modular, so can be fairly easily expanded to meet your increasing demands for more fresh vegetables. There is a lot of help and new ideas, which can be exchanged with fellow growers and manufacturers and their distributors by personal visits to stores (Fig. 125) or by the Internet. Hydroponic societies disseminate information through conferences, trade fairs, bulletins or newsletters as well as by the Internet. You should feel free to become part of these organizations and electronic communication as everyone has similar interests and are always looking for new methods and ideas to further advance the science of hydroponics. Enjoy yourself with the plant experience!

Figure 125. Display of various growing systems in a hydroponic store.
(Courtesy of American Agritech, Tempe, AZ).

Index

A

Adventitious roots, 20
AeroFlo^2, 102-106
AeroJet®, 107
Aeroponic systems, 79, 94-99, 107-113
A-Frame, 83, 89-92
A-Frame NFT, 89-92
Agrimek, 63
Air stones, 81
American Agritech, 82, 100, 107, 110-113, 122-125, 142, 154
American Hydroponics, 83-89, 92-94, 100, 142
Aphids, 60
AQ 10, 65
Aqua Farm, 115
Aquarium pump, 4
Asociacion Hidroponica Mexicana A.C., 147
Atomic ionic state, 66, 68
Australian Hydroponic & Greenhouse Assoc. (AHGA), 47
Automatic float valve, 28
Azatin, 63

B

Baby Bloomer, 100, 102
Ballast, 34
Bark, 80
Beneficials, 62
Beneficial microorganisms, 62
Bioagents, 8, 62-65
Blossom-end-rot (BER), 76, 77
Bok choy, 90, 132
Botanigard, 63
Bottled gas, 37
Bucket system, 1
Bug-scan cards, 60

C

Carbon dioxide, 37
Carbon dioxide generators, 37, 38
Caterpillars, 60
Catfacing, 77, 78

Centro de Investigacion de Hidroponia y Nutricion Mineral, 147
Channels (Gullies), 82-89
Chlorosis, 76
Cinnamite, 63
Circular light movers, 35, 36
City Green hydroponicum, 4
Clone Machines, 110
Coco coir, 4, 80, 103, 119, 122, 136-138
Column culture, 137
Conical reflectors, 30-33
Conversion lamp, 30
Crop changeover, 59, 60
CropKing, Inc., 142
Cucumbers, 10, 17, 18, 23, 27, 39, 41, 44, 48-53, 58, 59
  plant schedules, 58, 59
  plant spacing, 27
  seedling temperatures, 23
  training, 39, 41, 44, 48-53
Cultural practices, 26-65

D

Day neutral, 24
Deformation, 76, 77
De Ruiter Seed, Inc., 145
Diamond Lights, 100, 101, 142
Dipel, 62
Diseases, 60, 64, 65
Disinfectant, 59
Dissolved oxygen, 80, 81
Dormancy, 23
Dutch Bato Buckets, 119-122

E

Ebb & Flow, 92-94, 100, 101, 123, 125
EC meter, 75
Econojet®, 107, 122
Electrical conductance, 74
Electrical conductivity (EC), 74-76
Electric toothbrush, 53, 54
Essential elements, 66, 68, 76
Eve's Garden, 119-121
Everbearing (day neutral), 24
Excelsior wood fibers, 2
Expanded clay (Leca-clay pellets), 3, 80, 95, 101, 114, 125

F

Fertilizer salts, 68
Fixture, 34
Flat, 5
Float valve, 95
Foot candles, 24, 29
Fruit cracking, 77, 78
Fruit set, 55, 56
Fungus gnats, 60
Future Farms, 89, 90

G

Garden Fresh salads, 8
General Hydroponics, 100-106, 110, 114-121, 142
Generative state, 56
Gnatrol, 62

Government & Universities-
Hydroponics, 143, 144
Gravel, 1
Gravel culture, 80
Growing cubes
Growing tray
Grow cubes, 15-18
Growing tray, 3, 14, 15, 92, 101, 103, 105, 107, 108, 123-125
Grow tubes, 95-99
Grorox, 103, 104, 115, 119
Gullies (Channels), 82-89
Gummy stem blight, 21

H

Heavy Duty (HD) One- & Two-Tray Systems, 92
Herbs, 10, 11, 27, 57, 58, 101, 103, 128-131, 133, 135, 137, 139
Heydite shale rock, 3, 80
High intensity discharge (HID), 30
High pressure sodium (HPS), 30
Hobby Farm, 100, 101
Hobby hydroponics, 4
Homegrown Hydroponics, 100, 142
Horizontal reflectors, 31-33
Hydromall, 141, 143
Hydroponic forums, 148
Hydroponic magazines, 148, 149
Hydroponic Merchants Association (HMA), 147
Hydroponic organizations, 147
Hydroponic services, 147
Hydroponic Society of America (HSA), 147
Hydroponic supplies, 142
Hydroponics-definition, 79

I

Immobile elements, 76
Inorganic, 66
Integrated pest management (IPM), 62, 142
International Society for Soilless Culture (ISOSC), 147
Internet, 8, 141, 149, 153
Internet chat clubs, 147, 148
Inverted bottle, 6
Iron chelate (FeDTPA), 71
Irrigation cycles, 28

J

Jiffy-7 peat pellets, 15
Jetfilm®, 82
Jetflo Mini®, 124, 125
Jetstream®, 123, 124
Jetstream Mini®, 124, 125
Johnny's Selected Seeds, 145

K

Kocide 2000, 65

L

Leafminers, 60
Leca-clay pellets, 3, 80, 103, 104
Leggy, 18, 24, 29
Lettuce, 11, 12, 19, 23, 27, 56, 57, 82, 101-103, 132
  plant schedule, 56, 57
  plant spacing, 27
  seedling temperatures, 23
  transplanting, 19
Light, 24, 29
Lights, 30-26
Light Manufacturing Company, 32-36, 118, 119, 142
Light Movers, 35, 36
  circular, 35
  linear, 35
  Sun Twist, 35, 36
Litter tray, 2
Lux, 24, 29

M

Macroelements, 69, 70
Merck Color pHast, 73
Mesclun mix, 6
Metal halide (MH), 30-34
Microelements, 69, 71
Micro Gardens, 107, 110
Micronutrients, 69, 71
Millimhos (mMhos), 76
Mobile elements, 76
M-Pede, 63
Multi-pack trays, 14
Mylar, 33

N

Necrosis, 76
Neem-X, 63
New Mectin, 63
NFT, 18, 28, 80-92, 100, 103
NFT channels, 82
NFT Combo Gully Kit, 86
NFT Gully Kits, 83-85
NFT Rockwool Gully Kit, 84, 86
NFT Wall Garden, 86-88
North American Greenhouse Supplies, 94-99, 142
North American Hydroponics, 142
Nutrient reservoir *(See reservoir)*
Nutrient solution, 69, 71, 72
Nutritional disorders, 76-78

O

Oasis Horticubes, 15, 18
One- & Two-Tray Econo System, 101
Organic, 66, 67
Ornamental Edibles, 145

P

Parabolic reflectors, 31-33
Parasites , 62
Parts per million (ppm), 74
Peanut hulls, 2
Peppers, 12, 17, 23, 26, 27, 39, 40, 44, 47, 57
  plant schedules, 57
  plant spacing, 26, 27

seedling temperatures, 23
training, 39, 40, 44, 47
transplanting, 17
Percent germination, 16
Perlite, 3, 6, 14, 80, 114, 125, 126
Pesticide free, 67
Pests, 60, 144, 145
pH, 72-74
pH paper, 73
Phyton27, 64
Pipe Dreams, 94-99
Pipe Dreams Balcony18, 95, 96
Pipe Dreams Balcony32, 95, 97
Pipe Dreams-PD64, 95
Pipe Dreams-PD96, 98
Pipe Dreams-PD60, 95, 99
Planting schedules, 56-59
Plant nutrition, 66-78
PlantShield, 65
Plant spacing, 19-23, 26-28
Plant training, 38-53
P.L.Light Systems, 142
Plumbed Grow Trays, 123
Pollination, 53-56
Popular hydroponics, 1, 136-140
Potassium hydroxide, 74
Powdery mildew, 65
PowerGrower, 115, 117, 118
Practical Hydroponics & Greenhouses, 148
Predators, 62
Pruning, 42-53
Pyrethrins, 63

R

RainForest, 110
Raft culture, 18, 139
Receptive flowers, 54
Red-spider mites, 60
Reelenz reel support hook, 41
References, 149-151
Reflectors, 31-33
conical, 31, 32
horizontal, 31, 32
parabolic, 31-33
Relative humidity (RH), 54
Renewal umbrella training, 48, 49
Reservoir, 4, 28, 89, 91, 92, 100-107, 115, 122, 125, 134, 135, 138, 139
Rice hulls, 2, 80, 136
Richters Herbs, 145
Rijk Zwaan USA, 145
Rockwool, 15-18, 80, 114, 123-126
Rockwool blocks, 16-19, 81, 84, 123
Rockwool cubes, 15-17, 19, 84, 85, 101, 123
RootShield, 65

S

Sand, 67
Sawdust, 2
Seed houses, 145, 146
Seedling light, 24
Seedling temperatures, 23, 24
Seeds, 6, 9-17, 145, 146
Short-day, 24

Side shoots, 42, 43, 48, 50, 53
Singapore Society for Soilless Culture, 147
612 NFT Production Unit, 88, 89
Sodium hydroxide, 74
Soil, 67
Soilless culture, 79, 92, 100, 114
Soil organisms, 67
Sources of essential elements, 68, 69
Sowing, 14-17
Spintor, 63
Stoker Horticultural & Hydroponics Supplies, 142
Stokes Seeds Ltd., 145
Stock Solution, 69, 71
Straw, 2
Strawberries, 24, 25, 82, 88
Stringing, 39-41
Stunting, 76
Styrofoam boards, 18, 137, 139
Styrofoam pots, 126, 127
Subirrigation, 1
Substrates, 2, 14-18, 114
Sulfuric acid, 74
Sun Twist, 35, 36
Supplementary lighting, 8, 29-36
Support wires/hooks, 31
Symptoms, 76-78

T

Temperatures, 28, 29
Tendrils, 48, 49
Terrace Hydrogarden, 83
The Growing Edge, 149
The Living System, 118, 119
Thermograph, 29
Three-Stack System, 135
Thrips, 60
Tomahooks, 39, 40
Tomatoes, 13, 17, 23, 26, 53-57, 76-78
  plant schedules, 57
  plant spacing, 26
  pollination, 53-56
  seedling temperatures, 23
  training, 39-46
  transplanting, 17
Transplanting, 18-22
True leaves, 19-21
Truss hooks, 43-45
Two-Stack System, 133-134

V

Varieties, 9-14, 24, 25
V-cordon, 26, 41, 44, 48, 99, 125
Vegi-Table, 92-94
Vegetative state, 42
Vermiculite, 6, 14, 80
Vertalec, 62
Vertical Garden, 133-136
Vertical plant towers, 19, 133-136
Verti-Gro, Inc., 126, 133, 142
V-Frame, 94-99
Viability, 16
Vine clips, 41-43
Volcanic rock (pumice), 4, 80

W

Water analysis, 71, 72
Water culture, 2, 18, 79-113
WaterFarm, 115, 116
Watering, 28
Watts, 31, 33
Web sites, 8, 141-149
Whitefly, 60
Wood shavings, 2

X

Xentari, 62